T0227401

Revenue Assurance

Expert Opinions for Communications Providers

Revenue Assurance

Expert Opinions for Communications Providers

Eric Priezkalns

with David Leshem, Ashwin Menon, Hugh Roberts,
Güera Romo, Lee Scargall, David Stuart, Mike Willett,
and Mark Yelland

CRC Press
Taylor & Francis Group
Boca Raton   London   New York

CRC Press is an imprint of the
Taylor & Francis Group, an **Informa** business
AN AUERBACH BOOK

Auerbach Publications
Taylor & Francis Group
6000 Broken Sound Parkway NW, Suite 300
Boca Raton, FL 33487-2742

© 2011 by Taylor and Francis Group, LLC
Auerbach Publications is an imprint of Taylor & Francis Group, an Informa business

No claim to original U.S. Government works

International Standard Book Number: 978-1-4398-5150-0 (Hardback)

Visit the Taylor & Francis Web site at
http://www.taylorandfrancis.com

and the Auerbach Web site at
http://www.auerbach-publications.com

Contents

Figures

Figures

xiii

Tables

Foreword

I made the transition into the world of revenue assurance (RA) at the time before the gold rush that was to ensue in future years, at a time when most laypeople, myself included, struggled to even understand the meaning of the word "assurance," and at a time when the phrase "explosive growth" still applied to the vast majority of mobile telcos. Today, much of the landscape within which a successful RA function operates is almost unrecognizable from those humble beginnings, but the core disciplines, behaviors, and capabilities required back then remain paramount even today.

This book draws upon the actual experience of a number of RA professionals—those who have actually lived through the journey, learned from setbacks, and adapted their approaches to determine best practices in the field. This work isn't an academic exercise that will prove impossible to implement within your organization but one that is built upon the bedrock of actual implementation experience.

Spending many years in the RA arena, I have seen the model adapt and grow, the tools evolve, and the processes mature, but the most critical components of success have remained constant throughout—namely, the quality and experience of those individuals within the RA team itself. The world of RA has always been curiously collaborative. Experiences gained are rapidly shared, thought leadership is prized above commercial advantage, and productive teamwork is seen as vital within the RA function. The scope and depth of the understanding required to assure processes that touch most of the enterprise cannot be centralized into one all-knowing mind. Instead, teams have to synthesize individual strengths, knowledge, and experience to achieve a collective and comprehensive outcome.

Fortuitously, we recognized from the start that you had to build an expert team and accepted that we should pay for the best, avoiding the urge to capitulate to a "false economy" when given the option to save a few dollars and recruit a lesser individual. If you consider the fact that the business is constantly changing and inventing new and more complex ways to lose revenue inadvertently, you have to have an RA team that can adapt constantly to increasing complexity while retaining control over the existing processes. To avoid having to swap team members as the landscape gets more complex, you have to recruit and retain scalable individuals who retain their thirst and enthusiasm throughout.

This thirst and enthusiasm, while innate in a few, needs to be fostered and encouraged in most people, and as you inevitably reach a plateau on your journey, you will need to draw upon thinking bolstered from external sources that cannot be entirely grown organically. The odd paradox of RA is that ignorance and success are difficult to differentiate, and many functions are reducing their headcount on the back of diminishing amounts of demonstrable successes. I would argue that now is the time that RA functions to be even more inventive, more collaborative, and more disciplined than ever before. Many of the ingredients of loss are still out there and, in many regards, are more abundant today than in the past. Sure, you have to understand more and look a lot harder, but that's what makes this activity as compelling and rewarding as it has ever been.

In summary:

Recruit and retain the best people you can, without compromise.

Create and foster a team environment with constant collaboration and learning as a necessity of the operating rhythm.

Borrow from the rest of the RA community at every opportunity.

When pursuing your revenue assurance goals, I hope this book will be a vital source of inspiration and a catalyst for future thinking that will invigorate you into new areas of opportunity.

Julian Hebden
Director of Enterprise Information Management, Telstra
Former Director of Organisational Development for Information
Management Systems, HM Revenue & Customs
Former Vice President of Commercial Intelligence of T-Mobile
International Group
Former Head of Revenue Assurance, Billing and
Carrier Services, T-Mobile, UK

Acknowledgments

Thanks to Rich O'Hanley, CRC Press, for seeing the potential for this book without hesitation and for his straightforward helpfulness ever since. Thanks to Lucinda Steffen for insisting she would buy a copy though she does not work in revenue assurance or even in the communications industry. Thanks to Ian Elms and Adria Meserve for their hospitality, forbearance, and help during the final weeks of preparing the manuscript.

Thanks to everyone who reads *talkRA* on a regular basis and to everyone who has supported the site over the years; without you, there would be no Web site, and there would be no book.

Authors

Eric Priezkalns

Eric Priezkalns is the founder of *talkRA.com*, the Web site that gave rise to this book. He is currently an assistant director at Qtel International responsible for enterprise risk management. Eric has over 10 years of experience in the communications industry in the fields of risk management and revenue assurance.

Eric was the original (and much imitated) revenue assurance blogger. Having built a loyal readership worldwide, Eric decided to join forces with other thought leaders by forming *talkRA*, which went on to establish itself as *the* Web site for practitioners to get insightful views and the latest news about revenue assurance.

Eric previously worked as head of controls for Cable & Wireless Group; best practice manager in the revenue assurance, billing, and carrier services function of T-Mobile, UK; and the billing integrity manager for Worldcom, UK. He first worked as a consultant in the Enterprise Risk Services division of Deloittes, where he also qualified as a chartered accountant. Eric has a master's degree in information systems and a bachelor's degree in mathematics and philosophy.

Eric is very well known in international revenue assurance circles as a founding participant of the TM Forum's Revenue Assurance team, and he is now the leader of the TM Forum's Enterprise Risk Management team. He was the driving force behind the revenue assurance maturity model. He was one of the editors and contributors of the TM Forum's *Revenue Assurance Overview* and its *Revenue Assurance Guidebook*. It was in the first version of the TM Forum's *Overview* that Eric introduced the de facto standard definition of revenue assurance.

Eric was responsible for the program that saw T-Mobile, UK become the first CSP to satisfy the UK regulator's stringent metering and billing accuracy expectations across all communication services without exception, including those provided using the then-new 3G network.

David Stuart

David is the assistant director of revenue assurance and fraud management at Qtel International. A revenue assurance expert with over 10 years of practical experience within telecoms revenue assurance, David graduated from Brunel University with a degree in mathematics and statistics, after which he moved into the financial services sector.

David started his revenue assurance career at MCI WorldCom, where he designed and project-managed the development of a traffic assurance tool covering a dozen business units' fixed line traffic.

From here, David moved to One2One/T-Mobile, where he provided architectural assurance for all of the company's developments. David's expertise in the company's architecture led him to be made test manager for UK's first 2.5G launch.

From T-Mobile, David moved to a specialist RA consultancy, 2Helix, where he held the position of senior RA consultant. He worked on numerous projects, all delivering revenue assurance software, often custom-designed for the client's needs.

After 2Helix, David moved to Cable & Wireless International, where he was again responsible for delivering revenue assurance tools, this time to CWI's 33 business units.

Mike Willett

Mike Willett currently works as a consultant. He has over 12 years of experience specific to the telecommunications industry in the field of fraud and revenue assurance.

Until November 2009, Mike was the director for fraud and revenue assurance at Telstra Corporation, Ltd., Australia. Mike was at Telstra for 6.5 years and led the fraud and revenue assurance function in times of great organizational change as Telstra underwent its massive transformation program. His interest is in understanding theoretical approaches to improve revenue assurance outcomes and, more importantly, in how these can be practically implemented to provide tangible and recognizable business value.

Mike started his career at BellSouth (now Vodafone) in New Zealand and then moved to Praesidium Services in the United Kingdom. During this time, he had the opportunity to consult with a number of service providers and vendors around the world and see how revenue assurance was perceived and managed (and mismanaged) in a number of different operating and cultural environments.

Mike graduated from the University of Auckland in New Zealand with degrees in psychology and marketing.

Güera Romo

Güera has 13 years of experience in business transformation in the engineering, defense, government, banking, and telecommunication industries. She has experience in mergers and acquisitions, rightsizing, redeployment of personnel, business process reengineering, system selection, and implementation. Before this, she spent five years in finance and business administration. During this time, she was an accountant, after which she transitioned to financial application support on Oracle.

Since 1998, Güera has consulted in revenue assurance, billing, and customer care to two fixed-line and two mobile operators in South Africa and in the United States. At MTN, South Africa, she was responsible for establishing and managing a revenue assurance, fraud, and law enforcement function; sourcing an RA automation tool; and replacing a fraud management system.

Güera holds a BCom Hon (industrial and organizational psychology) degree and is currently pursuing a research master's focusing on the knowledge, skills, and abilities required to practically implement revenue assurance.

She is an independent consultant and an academic researcher.

David Leshem

David is an expert in telecom and utilities enterprise solutions—billing, profitability, business intelligence, customer retention, churn, and revenue assurance.

David has worked with major carriers all over the world, creating an enviable track record in improving the bottom line of telecom companies. He brings in-depth expertise to fixed-line, MNO, and MVNO businesses, helping them to get the best in pricing, margin reasonableness reconciliation, cost-effective customer retention and acquisition, and multiple revenue stream assurance.

David has international experience in addressing the financial challenges faced by telecom providers.

Ashwin Menon

Ashwin Menon began his foray into revenue assurance as an implementation on-site engineer with Subex. Being trained in both revenue assurance and fraud management, he undertook projects specifically pertaining to revenue assurance. During the course of his career, he has been involved with various clients across the Asia-Pacific region, including tier-1 telcos.

Ashwin has witnessed various leakage scenarios across telecom operators in the Asia-Pacific region, and has been privy to the different methods and controls that were implemented to target and plug issues in a telecom operator's value chain.

Ashwin is currently employed at Subex as a customer solutions consultant.

Hugh Roberts

Hugh Roberts is senior strategist for Patni Telecoms Consulting (formerly Logan Orviss International) working with network operators, service providers, VAS suppliers, and vendors to develop their service and product strategies, market and brand positioning, revenue fulfillment, business modeling, and the interpretation of new business requirements into technical and business architectures. Hugh is also a nonexecutive director of TeleOnto Technologies, an Indian start-up specializing in revenue assurance and business intelligence analytics. Formerly, he was development director of the TM Forum, where he was responsible for the introduction of the TM Forum's BSS Teams for revenue assurance, for content and data charging, and for pre–post convergence. He remains an active contributor to the TM Forum's technical reports and guide books.

Hugh is a frequent media commentator and analyst, speaker, and chair of industry conferences around the globe, as well as the author of numerous articles for the trade, national, and international press.

In his role as consultant director to IIR's telecoms division, Hugh advises on strategy for BSS, OSS, and revenue management development, and he has been associated with the major show *Billing & Information Management Systems* since 1994. He has been integrally involved in pioneering business-to-business Web and e-mail marketing and also has been chairman of the World BSS Awards Judging Panel since 1997.

Before this, Hugh worked in the entertainment sector, where he was responsible for bringing award-winning real-time interactive control and HMI systems to market for museums, theme parks, nightclubs, and major performance artists. He is an artist and graphic designer and has worked extensively in the music industry, including pioneering work on the development of three-dimensional sound and psychoacoustics.

Lee Scargall

Lee is director of enterprise risk management, with overall responsibility for revenue assurance, fraud, risk management, financial controls, and finance systems, at

Qtel International. Based at the head office in Qatar, Lee provides leadership and strategy for Qtel International's 16 business units, covering a combined subscriber base of 60 million people across the Middle East, North Africa, and Asia. Lee has over 12 years of experience in the telecommunications industry, having held previous positions as head of revenue assurance and risk management at Cable & Wireless and as business development manager at T-Mobile, UK. Lee has also been employed by Deloitte Consulting, working on projects for both mobile and fixed-line operators across Europe and the United States. Lee was awarded a PhD in electrical and electronic engineering from the University of Newcastle upon Tyne, UK, for advanced research in 3G video-telephony.

Mark Yelland

Mark is a revenue assurance consultant with over 10 years of experience around the world. He coauthored with David Sherick *Revenue Assurance for Service Providers*, the first new book on revenue assurance in five years, which was published in December 2009. He regularly contributes to discussions on LinkedIn, has had articles published via the Telecom Managers Forum, and has guest blogged on *talkRA*. His two passions are making RA more accessible to organizations feeling the pinch, and polar bears (see his Web site, www.raaiim.co.uk, for details).

Mark runs his own consultancy, raaiim ltd, formed after leaving THUS when it was taken over by Cable & Wireless, and is forging alliances with strategic partners such as Xintec SA.

Before becoming a consultant, Mark was head of revenue assurance at THUS for five years. During this time, he managed the successful implementation of a revenue assurance architecture system (based on SS7 signaling), worked with the teams to implement a new mediation system, and improved the quality of the retail billing. He helped THUS to gain approval against the Ofcom Metering and Billing scheme (2003). He was selected to be part of the experts working group that developed the 2008 scheme and demonstrated his confidence in the scheme by helping THUS become the first organization to be approved against the new standard.

Before joining THUS, he was director of revenue assurance for Cable & Wireless, providing expertise, training, and support to the different organizations around the globe.

He holds an engineering degree from Cambridge University and a master's in business administration from the Open University, and he has been a chartered engineer since 1976. This combination provides a useful balance in matching revenue assurance issues with requirements of businesses, allowing Mark to be able to operate in both technical and nontechnical areas with equal ease.

Abbreviations and Definitions

Term	Description
AAA	Authentication, Authorization, and Accounting
ACL	Audit Command Language
ADSL	Asymmetric Digital Subscriber Line
AMA	Automatic Message Accounting; modular CDR devised by Bellcore
ANSI	American National Standards Institute
A Number	The calling party's telephone number or CLI
AR	Accounts Receivable
B2B	Business to Business
BCM	Business Continuity Management
BI	Business Intelligence
B Number	The dialed telephone number
BPR	Business Process Reengineering
BSS	Business Support System
BV	Billing Verification
C&W	Cable & Wireless Plc
C7 CCS	ITU signaling system number 7
Capex	Capital expenditure
CC&B	Customer Care and Billing
CDR	Call Detail Record
CEO	Chief Executive Officer
CFO	Chief Financial Officer
CIO	Chief Information Officer
CLEC	Competitive Local Exchange Carrier; a term used mostly in the United States to describe network operators competing to provide local-loop connections

Term	Description
CLI	Calling Line Identifier
CMMI	Capability Maturity Model Integration; IT maturity framework devised by the Software Engineering Institute
CMO	Chief Marketing Officer
CMTS	Cable Modem Termination System
COO	Chief Operating Officer
COSO	Committee of Sponsoring Organizations of the Treadway Commission; private body that advises on risk and controls in business
COTS	Commercial Off-The-Shelf
CP	Communication Provider
CPE	Customer Premises Equipment; communications equipment located on the customer's premises
CPP	Calling Party Pays
CPU	Central Processing Unit
CRM	Customer Relationship Management
CS	Customer Services
CSP	Communication Service Provider
CSR	Customer Services Representative
CTO	Chief Technology Officer
CWI	Cable & Wireless International
DDI	Direct Dialing Inwards
DFD	Data Flow Diagram
DOCSIS	Data Over Cable Service Interface Specification; standard for high-speed data transfer over cable TV system
DSL	Digital Subscriber Line
DSLAM	Digital Subscriber Line Access Multiplexer; connects DSLs to the backbone
DSO	Days Sales Outstanding; measure of average time to collect debt
EBITDA	Earnings Before Interest, Taxation, Depreciation, and Amortization
Egress	Outgoing point at which calls or data leave the network
ELT	Extract-Load-Transform
ERM	Enterprise Risk Management
ETL	Extract-Transform-Load
eTOM	enhanced Telecom Operations Map; process map from the TM Forum and recommended by the ITU-T

Term	Description
EV	Expected Value
FMS	Fraud Management System
FTP	File Transfer Protocol; protocol for moving files from one machine to another in TCP/IP networks
Gbit/s	Gigabit (1000 million bits) per second
GPRS	General Packet Radio Service
GSM	Global System for Mobile communications
GUI	Graphical User Interface
HFC	Hybrid Fiber-Coaxial
HLR	Home Location Register
HMI	Human Machine Interface
HR	Human Resources
HTML	Hypertext Markup Language
ICT	Information and Communications Technologies
IDD	International Direct Dial
IFRS	International Financial Reporting Standard
IN	Intelligent Network
Ingress	Point where incoming calls and data pass to the network
IP	Internet Protocol
IPDR	IP Detail Record
IS	Information Systems
ISO	International Organization for Standardization
ISP	Internet Service Provider
IT	Information Technology
ITU	International Telecommunication Union; an agency of the UN
ITU-T	International Telecommunication Union – Telecommunication Standardization Sector
IVR	Interactive Voice Response
Kbit/s	Kilobit (1000 bits) per second
KPI	Key Performance Indicator
KPV	Key Performance Vector
KQI	Key Quality Indicator
Local loop	The connection between the customer's premises and the local exchange
M&A	Mergers and Acquisitions
Mbit/s	Megabit (1 million bits) per second
MIS	Management Information System
Modem	Modulator/Demodulator

Term	Description
MOU	Minutes of Use
MPP	Mobile Party Pays
MSC	Mobile Switching Center
MSISDN (number)	Mobile Subscriber Integrated Services Digital Network number; a unique identifier for mobile users
NGOSS	New Generation Operations Systems and Software; communication provider framework from the TM Forum
Opco	Operating company
Opex	Operating expenditure
OSI	Open System Interconnection model
OSS	Operational Support System
PDCA	Plan-Do-Check-Act
PDSA	Plan-Do-Study-Act
P/E Ratio	Price–Earnings Ratio
POTS	Plain Old Telephone Service; refers to basic analogue voice telephony
PRS	Premium Rate Service
PSTN	Public Switched Telephone Network
QoS	Quality of Service
RA	Revenue Assurance
RADIUS	Remote Authentication Dial-In User Service
RAID	Redundant Array of Inexpensive Disks
RDMS	Relational Database Management System
RFI	Request for Information
RFP	Request for Proposal
RM	Revenue Management
ROI	Return on Investment
SCP	Service Control Point; an element of the IN
SGSN	Serving GPRS Support Node
SID	Shared Information and Data model; data model of the TM Forum
SIM	Subscriber Identity Module
SIP	Session Initiation Protocol
SLA	Service Level Agreement
SME	Subject Matter Expert
SMS	Short Message Service
SNMP	Simple Network Management Protocol
Softswitch	Nonproprietary IP switch

Term	Description
Term	*Description*
SOX	Sarbanes–Oxley Act; reporting and internal controls obligation for some US-listed companies
SP	Service Provider (see also "CSP")
SQL	Structured Query Language; ANSI standard for interrogating and updating any compliant RDMS
SS7	ITU Signaling System 7
SSP	Service Switching Point; element of an IN
SWOT	Strengths–Weaknesses–Opportunities–Threats analysis
TAP	Transfer Account Procedure
TCO	Total Cost of Ownership
TCP	Transmission Control Protocol
TMF	TeleManagement Forum (see also "TM Forum")
TM Forum	TeleManagement Forum (see also "TMF")
Traffic	Collective description for the transmission of individual telephone calls
Transit	Any call passed from one carrier to a final carrier via a third party
UGC	User-Generated Content
USP	Unique Selling Proposition (or Unique Selling Point)
VAT	Value-Added Tax
VoIP	Voice over IP
VoWiMAX	Voice over WiMAX
WiMAX	Worldwide Interoperability for Microwave Access; wireless broadband access technology
X.25	Protocol for packet switched data transmission
XDR (or xDR)	Any kind of Detail Record

About This Book

Why We Wrote It

Eric Priezkalns

This book represents a new chapter of the *talkRA* project, an initiative dedicated to improving the discipline of revenue assurance (RA) for communication providers. The project started with a Web site that brought together bloggers from around the world, all experts in the field of RA. Over the years, a special dynamic has evolved, with the authors bouncing ideas off each other. Their debates are followed by a growing readership around the globe. The *talkRA* authors share their advice and opinions for free, motivated by a desire to develop the subject they feel passionately about. They believe that the best contribution they can make at this stage in the development of RA is to explore options for how it might be performed, not to assert a premature orthodoxy that would limit its growth. Exploration comes from completely open debate between thought leaders. With the *talkRA* project, there are no membership fees, no hard sale of training courses, and no censorship of differences of opinion. There is just an earnest desire to help RA become all it can be. With this book, the *talkRA* project turns a new page.

Because good debate is drowned out if too many try to speak at once, *talkRA* is not an open forum. Authors are invited to join on the basis of their personal track records in pushing the boundaries of knowledge and freely sharing their insights with peers. To be a *talkRA* author, one must be more than just an expert, although all of them are eminently qualified and very experienced. The *talkRA* authors share a philosophical belief that willingness to share and the unconditional freedom to debate their differences of opinion are vital to developing their craft. We also believe that no one person can be an expert in all aspects of RA. Our's is a hybrid discipline; even the best practitioners will inevitably have their relative strengths and weaknesses. For the authors themselves, *talkRA* also represents a learning opportunity, as they benefit from contact with peers who have different skill sets and backgrounds from their own. Within this book itself, you will sometimes see different authors taking different points of view on the same subject. We are unapologetic about this. In fact, we encourage it. By presenting all points of view, we believe we give more helpful advice to the reader, who can get a deeper understanding of the topic by contrasting how different members of the *talkRA* team would tackle the same challenge. However, irrespective of our differences, everyone in the *talkRA* team has the highest respect for his or her peers. We sincerely believe that we represent a collective effort to develop RA that is unique in the industry.

The book is presented as a series of articles from all of the *talkRA* authors. The articles are grouped into themed chapters. Do not expect a manual, as we simply do not believe it is possible to write a one-size-fits-all instruction book on how to do RA. If you feel differently, we recommend you switch off your brain and buy some software to do RA; machines are good at doing repetitive tasks the same way every time, whereas people are good for handling the variety of (often unexpected) challenges that RA throws at them. We do not deny the variety. We revel in it. That is why we wrote this assuming you already knew the difference between an SGSN and SQL, and that you are looking for higher-power ideas on how to tackle the challenge of RA and move up the scale from basic checks to optimizing performance. It is the ability to think outside of the box that distinguishes the real RA professional from the person who just happens to have "revenue assurance" in his or her job title.

Give a man a fish, and he is fed for a day. Give him a fishing rod, and he can feed himself for a lifetime. What, then, if we teach a man to make his own fishing rod? We aim to teach RA practitioners the art of teaching themselves how to do RA. This way, you will be able to adapt to every challenge you face. We will give you stimuli for your imagination and strategies for building RA capability, as well as practical guidance based on experience.

All opinions represented are those of the individual authors and not of their employers or of any industry body. They are not even the opinions of *talkRA*, as *talkRA* has no official opinion.

The *talkRA* team is the coming together of multiple insights to a common purpose, but do not take my word for it. One of my colleagues illustrates the point in the next article, as she gives her own take on the *talkRA* mission.

Is This the Wrong Question?

Güera Romo

I was speaking to the new RA manager at a previous employer and was thrilled to hear that my friend had taken over that department. The well-being of those team members has been renting space in mind for some time. I e-mailed an ex-colleague to say "Thanks for looking out for what remains of this team" and was struck by a thought: What exactly was renting space in my mind? The well-being of the function I am passionate about, or the great many people putting heart and soul into it?

The mind and language philosopher Wittgenstein said that language belongs to groups, not to individual or isolated minds. Language reflects communal practices, and specifically, how a community uses words in a language. The language also has context and is infused with sociocultural detail, which further informs the common understanding and adoption of the specific language as practiced.

The term "revenue assurance" has a generic meaning to all who are interested in the subject, yet not specific enough to categorize its components when we reduce the term to a cipher, or a bit, or an atom. A lot of what is going on in blogs, advertorials, vendor white papers, and certification efforts is based on pretty much the technical stuff—the how to, or with what cool tool. The sales pitch is about the technology, with added benefits of consulting and on-site support. It assumes that the question is "*how to do RA?*"

Is this the question? Does answering this question inform of us of what exactly RA is? I could not help but reflect that, for many of us in the RA industry, RA is a logical certainty, one for which we are prepared to fight, motivate, and convince. Is the question, perhaps, *whether to assure revenue?* The answer would certainly be *"yes."* Does this require a dedicated team and specialized tools? Ask a commercial bank or car manufacturer, and the answer is *"no."*

Is RA, perhaps, rather a personal relationship? In some of the material I review for the academic research I do, I find reference to techniques to convince the CFO of the merits and benefits of investing in RA. The mere fact that there are executives who do not immediately see the "must have" of this function and who need a compelling business case to assign funds in *this* direction (rather *than* that direction) is good cause to revisit the question.

What is the correct RA question to ask to get an answer of 42?

Chapter 1

Beginning at the Beginning

Ten Tips for Starting Revenue Assurance

Eric Priezkalns

One of the most common questions I am asked is "how do you start doing revenue assurance (RA) in a communication provider that has never done any before?" I suppose I always put off writing a public answer because I get paid for providing consultation. Giving away the answers for free means I get less money! However, here are my 10 tips for how to start doing RA for the first time:

1. *Use a list of major variances to outline ambitions, and then use a list of detailed leakages to further define scope and priorities.* Setting a scope and communicating what that scope is can be one of the biggest impediments to getting forward momentum when establishing RA for the first time. Begin by establishing your high-level ambitions using the variance analysis described in the article entitled *Using Variances to Set the Scope for Assurance* in Chapter 4. Most practitioners begin with only a narrow scope of finding and resolving core leakages, but not all do so. Depending on the communication provider's business model and the priorities communicated by the executives, it may be appropriate to include areas such as cash collection, cost management, and network utilization in the scope from the very beginning. If so, be clear about this.

    Once the broad ambitions have been set, it is time to break out the specific goals. In practice, many new RA functions find they cannot cover the

1

full scope they set for themselves with the resources available, so they need to have a clear sense of priorities. Decide what detailed controls you want to ensure are in place during your first year and maybe for a couple of years after that. Do not bother to forecast after that; the business will keep changing and you can only guesstimate how quick your progress will be as you cannot know the scale of challenges and leakages in advance. Do not try to be clever by trying to work out every possible leakage your business might suffer. Any list you come up with is bound to be incomplete, so save yourself some time and trouble and borrow from the TM Forum's leakage framework. You can find it in the TM Forum's *RA Guidebook GB941* (Priezkalns 2009). Not every leakage in the list will be relevant to your communication provider. Identify which ones are relevant, and then make some decisions on which ones will be in scope and which ones will be outside the scope. Give yourself a sense of priority as to which you want to investigate and deal with first. By doing this simple task, you have a solid foundation for explaining your role to your bosses and other people in the business and for showing how you are managing the work by identifying what to cover now and what to leave for later.

2. *Gather relevant information already known to someone in the business.* A lot of RA is good common sense about how to run a business and stop mistakes. This means that even in a business where nobody has been employed to do RA, there probably are people who already do the relevant tasks. They may be monitoring a specific leak or doing work that helps prevent some kinds of leaks. Instead of duplicating their work in ignorance, make the effort to speak to people around the business and ask if they have measures that will help fill the grid of leakages you created from the leakage framework. Perhaps billing operations do some sample checks of bills. Perhaps internal audit did an audit of suspense management last year. Perhaps the engineers make test calls that are compared to bills. Perhaps internal departments check internal phone costs, and this highlights occasional problems. Perhaps customer services keep track of the number of upheld complaints about billing accuracy. In short, look through your grid of the leaks to cover and identify if there is somebody in the business who might have some useful data. Find them and speak to them. If there is some data, even if not perfect, it will help you to determine if there is a serious problem or if the risk of leakage is under control. Perhaps you can generate some quick wins by suggesting small changes to the tasks people already do. For example, perhaps the Billing department does sample checks of bills, but they do not keep a record of the value of errors they find. Persuading them to keep a record will give you a new metric. If, after speaking to people, you find there is no data for some of the leakage points, then you have a stronger business case for implementing new measures and controls to fill the gap because you can be confident that nobody

else is managing the risk. Finally, the other big benefit of talking widely about RA is that you will get a good idea of who are your natural allies in the business, and hence who will work with you to deliver results.

3. *Prove your worth with quick wins.* After you have done your survey of who knows what, focus your efforts on one or two areas that are very likely to generate straightforward and valuable financial benefits. Even if you would prefer to do something else first, pick the areas where you are very confident there are leaks with a significant financial value, where you have a good idea how to stop them, and where you believe you can get results quickly without needing to spend much money. Favor issues where you feel that other people in the business are more likely to support and help you, and avoid those issues where you expect to face a lot of internal resistance. It is more important that you can show you can fix the leak, and demonstrate the value, than to pick the biggest leak. Prioritize smaller leaks if you can resolve them and show the benefits more quickly. If you can, also favor those leaks where you can recover some of the loss (e.g., through back-billing). Your focus must be on overcoming skepticism. Generate the belief that RA adds value to the business, which is why a smaller but quicker win may be better than a larger win that takes longer to deliver. Building confidence and a track record of delivery at an early stage will make it easier to overcome obstacles later on.

4. *Spread your coverage with "quick and dirty" audits.* Despite the advice to focus on quick wins, avoid the temptation to put all your resources into fixing known problems. Use some resources just to get some understanding of leaks that have not been covered before and where you have no data. A quick and dirty audit, perhaps based on a small sample, is sufficient to provide you with an impression as to whether an area has serious problems or is generally fine. Do not worry about following a perfect approach; just do a rough check to see if there are serious issues. If everything appears fine, you can report to management that you have done a little work and there is no urgent need to act, although you will need to do more detailed work later. If there are problems, you may find other opportunities for quick wins that you can prioritize above some of the leaks you knew about before.

5. *Emphasize versatility and a range of skills amongst your people.* The range of skills needed to do RA is phenomenal. It could involve everything from writing an SQL query to having an intelligent conversation about the IFRS for revenue recognition. It needs people who can understand technology and people who are good communicators. In the early days of doing RA, it will be very hard to assemble any kind of team, whether a project team, dedicated staff, or a virtual team, with the full range of skills needed. Be aware of this, and try to find individuals who are versatile, adaptable, and willing to take on

tasks that will move them out of their comfort zones. Do not employ people who feel they know it all already; prefer people who like to learn and expand their horizons. If you employ a programmer, try to find one with strong interpersonal skills. If you recruit an accountant, choose one who understands technology. Encourage team members to broaden their skills by learning from each other and by substituting for each other.

There are some very bad role profiles circulating around RA circles that seem to say that everybody in the team should have the same kinds of skills as each other. They make no mention of other relevant and useful skills. There are also some very silly profiles that ask for every imaginable skill to an unimaginable degree of proficiency. Be realistic and try to find the right blend of people who will complement each other. Most important, the manager must resist the temptation to recruit only those people from the same background as he or she is. It is harder for a manager to lead people with specialist skills the manager lacks, but good RA demands it.

6. *Admit your ignorance and ask stupid questions.* Every culture is different, so it is hard to generalize about this recommendation. However, RA people usually do better if they are seen as friends who work with the business, and not as policemen who hand out punishments instead of help. People in RA cannot know all the details, and the best way to learn is to talk to people in the company who know how things work and to learn from them by asking questions. This serves two important purposes. First, it educates you and your people. By understanding how systems and processes work from end to end, you may spot potential leakages, or opportunities for controls, that have previously been missed. Second, by asking a question you may make somebody else think again and realize there is a problem they had not thought of before. Perhaps an engineer is doing a very good job, keeping his boss happy and meeting his deadlines, but nobody asked him about possible financial implications if such-and-such happens. If you ask him, he may realize for the first time that the business is losing money and he may also be able to suggest how to do things differently. He may do this quicker than you could do it. When this happens, try to have a positive relationship and to share credit for finding and solving the problem. If you share credit, people will be willing to work with you in future. If you give blame or try to keep all the credit, then people will be discouraged from helping you again. It is important that your bosses also think of RA as a means of delivering the desired results by taking simple initiatives like talking to people and asking questions, as well as by involving lots of software and tools.

Remember that there are no "stupid" questions. Even a very simple question may highlight a misunderstanding or a mistake that is costing the business money. Be humble and appreciate that you can also make mistakes and misunderstand things, so find ways to confirm if your understanding is correct.

A straightforward way to do this is to explain back what you have learned. Be brave and ask questions, even when you deal with unpleasant people who think you are ridiculous to ask them. You will not convince everyone, but if you persist, you will succeed, and it is important for you to emphasize that in this world, in every company and in every country, mistakes will happen. It is your job to ask questions to try to identify and prevent those mistakes.

7. *Engage a broad range of executives with tailored messages.* Presentation is very important. If you present useful information, then you have the chance to become friendly and win the support of many executives, and not just your boss. If you present big totals but with no drill down, or long lists where only a few items are relevant to the person listening, you will lose support. If you present a good-sounding idea but no numbers, you will lose interest. Wait until you have some of your leakage map filled in. Then ask politely to arrange one-to-one meetings with a wide range of executives in addition to the executive you report to. If you have relevant information, you might end up with individual meetings with the CEO, CFO, COO, CIO, CTO, CMO, and so on.

   Call big roundtable meetings sparingly. If you talk about RA to a large and varied audience it may save you time but at the price of wasting the time of everyone else. Try to set agendas that concern most participants most of the time and show your appreciation for the time everyone contributes when they attend your meeting. Keep meetings with executives short and focus only on topics of relevance to the specific executive, while restricting the content to explaining a few big issues. Highlight the top few leakage numbers with a known cause, plus a few big risks areas where you have no measures and where there is no indicative data. Be clear about actions that need to be performed and try to schedule follow-up meetings that will discuss whether these did happen. In short, if you can show executives how you support them to meet their targets and how to look better when dealing with the rest of the executive team, you will acquire the influence to get information and to promote change all over the business.

8. *Adopt the mantra of "who, what, when."* Because of the variety of work in RA, you need very adaptable people to do it, especially in the early days. Even if you pick adaptable people, it is still natural for them to concentrate on the things they are good at, have experience of, and like to do. They may consciously or subconsciously do more of the work they like and not enough of other work that is more important. For example, some people will like to query data and detect problems, but do they follow through on the fix that needs to take place? Perhaps some members of the team like to find problems by creating process maps and holding workshops, but do they interrogate data to back up their theory by quantifying the actual amounts

lost? You must constantly strive to finish every job because there is a natural inclination to leave RA jobs half-done. Obstacles to progress can arise at any time; you must create a working culture where obstacles are overcome instead of allowing people to give up and try something different. Deal with this problem by adopting a simple mantra that everybody will repeat and will get used to following habitually. After every task, the next question on everybody's mind should be who needs to do what and by when to follow up. Only when the problem is fixed permanently can you stop asking who, what, and when.

The "who, what, when" mantra is useful to every team at every level of maturity. Get a head start by adopting the mantra at the very beginning and making it part of the team ethos. That way, when the next step is far from obvious, people still feel the responsibility of making the next step happen. Sometimes you know what needs to be done, but not who in the business should or could do it. Sometimes you know there is a problem, but are not sure what needs to be done to resolve it. Sometimes you know both who and what, but are not clear on the relative priority and urgency. Repeat the "who, what, when" mantra and encourage everybody you work with to take personal responsibility for finding the right person, defining the tasks that need to be done, and quantifying importance. In short, make it clear that it is everybody's responsibility to learn and ask questions to get answers, whether the answer is a name, a task, or the evidence needed to determine a relative priority. There will be very many questions to ask and very many answers you need to find. Encouraging people to take responsibility, to be autonomous, and to use their initiative is vital to rapid delivery.

9. *Adopt and adapt an off-the-shelf strategy by using the maturity model.* The Revenue Assurance Maturity Model, described later, can be used to help strategic thinking about RA. It can be used when just starting out on the road to RA or in a business with extensive experience of putting RA into practice. In the early days, strategy may not seem important but the self-assessment is quick and powerful; it can be performed within a few hours and requires no specialist knowledge or data. Though quick, the maturity questionnaire distils the collective experiences of many businesses. If you have walked around the business and created your leakage grid, and attempted to arrange a few meetings with executives, you will know everything you need to know to assess the RA maturity of your business. The results may surprise; just because you are starting RA does not mean the business is starting from the bottom in all aspects of what is needed. Using the maturity model will give you a simple basis for communicating strategic priorities to senior people and a quick mechanism to lay out a road map for how RA should develop. It ensures that the key pillars for good RA are all considered: organization, people, influence, tools, and processes. Making sure you think about all of

them will help you to balance priorities and ensure that you improve each aspect at the same rate. Using the maturity assessment is particularly helpful for avoiding traps and mistakes that occur sometimes, where an RA team makes rapid progress but then loses a sense of direction because it focused on too narrow a sense of its purpose.

10. *Get into the loop and get information about the future.* In the rush to fix the current problems with your business, it is easy to fall behind. Many RA teams discover, much to their horror, that they end up dealing with yesterday's problems, while paying insufficient attention to the problems of tomorrow. It can feel like the RA team is always struggling to play catch-up. Avoid this by pushing for the RA team to be represented, or at least informed, in case of any new system changes, business transformations, and product launches. You may not have the resource for a thorough review, but at least perform a high-level analysis of the potential risks for leakage and the key controls that should be put in place. Knowledge of what lies ahead will often change the priorities for RA and will help ensure that the most important considerations get the bulk of the limited resources available. A few simple questions about potential leakage also help the business because they can often lead to improvements in system and process design that will reduce the risk of errors. In addition, it is often much cheaper to implement controls if they are included during the initial design stage of a project.

Conclusion

These are my top 10 tips for starting RA. There are, of course, many other things that need to be considered when starting RA. It was a struggle to keep the list to just 10 tips. The beginning is a vital time, and the impression you create and how much momentum you generate will determine how rapidly you make progress for years to come. I know of big telcos that claim to have mature RA but where it is clear they started down the wrong path at an early stage. They may find it difficult to get back on track later, no matter how much is spent on staff or tools. This is because RA is much more than the job done by the RA department, and it is crucial to influence the rest of the business to behave in ways that support and enable the goals of RA. Funnily, the best time to influence the rest of the business is at the very beginning. At the beginning, there are no clear expectations and the lack of resources makes it easy for RA people to ask for help and show how they can work well with other staff. Make the wrong start, and other staff may fear the RA team, obstruct its work, or compete with it. It is harder to reverse the damage of bad relationships than to get the relationship right at the very beginning.

If you are starting RA for the first time, you have my sympathies because the challenge may be daunting. But follow these top 10 tips, and know that if you start right you can make a very important contribution not just to the bottom line but to how your business works. The beauty of RA is that when it is done right, the benefits are enjoyed in the short term *and* the long term. It cuts across many aspects of the business and provides a never-ending learning experience with a fresh challenge every day. That makes my eleventh tip the best of all: try to enjoy it!

Three Basics

Lee Scargall

If you are new to revenue assurance and are looking for some help, here are three suggestions to put on your "to do" list. I have used them time and again, and there is no quicker way to generate momentum than by leveraging this best practice guidance.

1. Use the framework in the revenue assurance maturity model to assess the maturity of business activities to deliver RA objectives in your organization. Pull together an action plan to progress to a higher level. The maturity model is described later, in the article "Exploring the Concept of Maturity in the Performance of Revenue Assurance" in Chapter 4.
2. Go through the inventory of leakage points in the TM Forum's *Revenue Assurance Guide Book* (Priezkalns 2009) to identify gaps and weaknesses in controls. This list can also be used to help standardize revenue assurance reporting across a multinational organization. In an article in Chapter 5, we will discuss a useful case study about how Cable & Wireless built the RA reporting for its international group.
3. Set yourself targets using the KPIs given in Annex A of the TM Forum's *Revenue Assurance Guide Book* (Priezkalns 2009).

This will not give you everything you need to put RA into effect but it will help you to construct a solid framework for everything else you do. Use this framework to structure your program of improvements.

Revenue Optimization for Greenfields

Ashwin Menon

Assuming that we had an opportunity to help a new operator (a "greenfield" operator) build a revenue assurance (RA) and fraud management practice, what would we inculcate into the "DNA" of the operator from day one? The experiment is directly focused on checking the impact of "early-in-the-day" enabling of RA and fraud management and on the extent to which the move would help the operator save on large leakage issues later on. In my opinion, the critical step for the operator would be to design and decide the RA and fraud management framework. The framework should encompass all aspects of reporting, tracking, the evolution path, and integration. Key points regarding a RA and fraud management practice from day one would be

(a) Planned integration of new network elements with RA as a test bed for system accuracy
(b) Building up of effective usage patterns for identification of fraudsters via deviant usage tracking
(c) Proactive verification of subscription data flow and subsystems
(d) Ensuring a "leak-proof" workflow to handle all issues, including rectification tracking

Of course, there are many more reasons for a new operator to take advantage of a fully fledged RA and fraud management function. One of the most important is that fraudsters who have attacked and possibly been ejected from other providers will typically find a greenfield operator to be a "soft" target. Another is that RA, simply be asking questions about how things work, helps to ensure that vital decisions are understood by the people making them. While keeping in mind that fraud management and RA might not be a key component of a new operator's rollout strategy, it is also important to appreciate that prevention is definitely better than cure.

It is absolutely clear to me that having a strong RA and fraud management framework in place from day one would definitely help an operator in both the short term and the long term. The progressive growth of the network will help the analysts to have a clear understanding of concern areas, as well as build a considerable in-house knowledge base. As a direct result of forward planning, the in-house teams would also have to implement strong workflows. Simple errors, like business document version errors, which can lead to contract disputes and leakage, could be foreseen and nipped in the bud.

Chapter 2

Revisiting the Foundations of Revenue Assurance

Revenue Assurance at the Crossroads

Hugh Roberts

To a great extent, revenue assurance (RA) in communication providers was born of the last recession.

Embattled CEOs were finally forced to come clean about the parlous state of internal systems that had expensively been put in place over the previous 10 years. Typically, even those that were fully functional were no longer fit for the purpose, and many intra- and interdepartmental processes were broken. Consequently, RA started its life with a boom—"low hanging fruit" were everywhere, and demonstrating the business case for any RA-related activity, given even the smallest sign of executive championship, was relative child's play.

As we enter a new economic cycle, RA has consistently demonstrated its worth, but in many operating environments (although certainly not all) the opportunities for easy revenue wins have been much harder to come by. Consequently, while RA has, by and large, retained its primary focus, the temptations for expansion and scope creep have been great.

In the interim period, there have been a number of external factors that have had a significant impact on the RA operating environment.

The first has been the drive toward ensuring shareholder value (sometimes even at the expense of business value) through increasing financial governance, ethical, business, and compliance requirements. While on the one hand, this has thrown into sharp relief the need for the business to focus on processes and risk, which has, in turn, undoubtedly bolstered awareness of RA, on the other hand, it has not necessarily done RA any favors. In some cases it has subverted RA activity away from its prime directive and into less strategically significant areas such as process management and compliance enablement.

The second has been the trend—both within the communication service provider (CSP) community and the supply chain—for aggregation through mergers and acquisitions (M&A). This, too, has had an upside and a downside for RA. Group-wide systems environments have again become more complex and the opportunities for RA rationalization with centralized coordination, resources, and power have increased, but at the same time the opportunity for the group headquarters to leverage economies of scale, impose "preferred supplier lists," and demand headcount reductions across its opco dominions is increasingly being aided and abetted by "RA justification." While efficiencies can undoubtedly be achieved, and in many cases are both necessary and long overdue, the potential impact on RA as it becomes ensnared in group/opco politics is that by having its interdepartmental communications skills and connectivity exploited, RA will be returned to an environment where it is treated by the opco business units with suspicion. RA will have become the harbinger of bad news rather than the provider of mission critical assistance in delivering operational effectiveness. Ultimately, if this approach continues without adequate safeguarding, not only RA's "client" relationships with target business units but also its bilateral relationships with closely coupled functions such as fraud management, internal audit and risk, will suffer.

The third trend, largely in response to the first two, has been the increasing effort to standardize and quantify all aspects of business and operational processes. Again, the monitoring and risk assessment of business effectiveness has to be a good thing, and for many operators who started with relatively immature business reporting capabilities, the ability to "actually know what is going on in their business" has been a revelation. However, these advances also bring with them the potential for business ossification and a tendency for "death by KPIs" in all of its various forms. Not only are operating environments in some CSPs becoming overburdened with quantitative metrics that are inhibiting the potential for very necessary business transformation in the light of changing new generation requirements, but important projects required to deliver new functionality are being inhibited even from consideration in case they conflict with reference architecture models that have become erroneously set in stone.

The fourth trend has been a response to the perceived success of RA itself, and indeed to RA's own attempts to expand its scope. Having successfully secured a degree of proactive control over leakage across a wide range of operational areas, many RA professionals—particularly those in more mature markets—have looked beyond RA toward a more strategic role with a broader level of influence on business operations. Areas such as input to, and even sign off on, new product development; increasing engagement in marketing and sales activities; and the development of consistent approaches to third party management and revenue share settlement are all coming within the remit of this expanded RA sphere of relevance. Moreover, given the huge variability in market penetration and maturity, customer expectations and regulatory responsiveness, together with the status of internal legacy systems and processes experienced by CSPs around the globe, the concept of "best practice" as a determination of implementation suitability and success factors is becoming increasingly harder to justify. While on the one hand, best practice is a useful benchmark for CSPs to aim at, as a reflection of how a specific function can be best optimized, on the other, the fact that analysts point at a particular solution as "the way to go" does not necessarily mean it will be optimum in every operating (and legacy) environment. Most certainly it does not necessarily guarantee optimized ROI, either in the short or in the long term. Without a doubt "best fit" has become a much better guiding principle, but this is and will always remain a subjective judgment call. One size definitely does not fit all.

A further problem in this respect has been the levels of understanding at senior management level (and elsewhere within the business) about the nature of standards themselves, and the adoption of an attitude that has assumed that technical process standards developed at the network layer and the business process standards developed at the IT and business layers are methodologically the same, and that they can be treated, managed, and developed in the same way. This is clearly not the case, and as a consequence—the sterling efforts of the TMF RA Working Group notwithstanding—the overall status of standards evolution within the BSS domain in particular remains something of a mess.

Billing is a good example. As a technical function there are technical standards for device interfaces that can be adopted; there are also process standards that optimize the implementation of these interfaces. However, billing processes also encompass a wide range of business activities that are dependent on intangible nontechnical factors—particularly those that affect customer interaction and marketing. In these areas billing policy is as closely related to corporate positioning and brand management as it is to the underlying platforms on which these processes are enacted. Technical disciplines can be very effectively managed by technical process standards. However, assuming that profitability can be guaranteed by the application of technology-based billing processes to the wider domain of revenue management as a business operation is likely to end in tears.

Unfortunately, at the same time, almost every software vendor across the OSS and BSS domains has laid claim to offering complete or near-complete RA capabilities within their product offerings, often under the banner of revenue management. While some of these offerings are genuine (and the need to embed core RA functionality into all operational systems is becoming a necessity for new generation product management and risk amortization), many are not, and most are primarily designed to reposition the functional set of their product suites on offer across more "strategically significant" dimensions. In addition, of course, all of the usual "guardian of the gatekeeper?" questions still arise. From a pure RA perspective, this muddying of the waters is not helping.

* * *

The big three areas of RA activity with highest visibility still remain:

1. Switch-to-bill/order-to-cash (CDR reconciliation/rating verification, etc.), which has the highest historic visibility in Europe
2. Inventory management, which has the highest historic visibility in the United States
3. Interconnection, which has the highest historic visibility in high-growth markets

To these has been added a fourth key element:

4. Analytics and Business Intelligence support

… And these listed areas remain those within which the "quickest wins" will continue to rise to the surface most easily. However, as a result of the increasing diversification of RA activity, the championship and sponsorship of RA is becoming yet more heterogeneous, with the heads of finance, audit, risk, IT, operations, and now marketing all having a vested interest in potentially maintaining strong controlling links over the RA domain. This is because of RA's access to a broad range of business information and its increasing power as a justificatory business mechanism (e.g., with its power to sign off on new products). Inevitably, both methodological conflicts and turf wars will increasingly embroil RA as a result (even in cases where RA might have no direct play).

This is the crossroads at which RA finds itself. Great opportunities exist for the advancement of both RA and the positive impact that RA professionals can make on the business abound. At the same time, there is a growing risk that RA will become embroiled in the changing political landscape driven by factors in both the market and economy.

Of course, politics is not the only SWOT factor on the horizon. Within the backdrop of ongoing business transformation, the repurposing of existing functionalities, development of new technologies, and new skills requirements are all continuously evolving. Here are just some of the new challenges:

■ *The growing importance of wholesale, and all that this will bring with it:* This is not just an extension of existing interconnect assurance; new factors include wholesale marketing considerations, third party settlement complications, revenue share complications, distributed rights management, UGC (user-generated content) management, attention data, integration with transit and peering management, end-to-end contract (document) assurance, real time inter-business data management complications, and the impact of multiple hostile audits that will inevitably follow in the wake of a more competitive wholesale environment.

■ *An enhanced service support environment:* This includes real time policy and data management complications, active mediation, exception charging, meta-data and algorithmic service enactment, increasingly complex customer segmentation and analytics, customer experience/interaction management with augmented personalization, and self-care complications (all required on a "time-to-market yesterday!" basis).

■ *The Internet:* While RA lessons learned from the unconverged mobile and PSTN environments will still be applicable, IP is both qualitatively and quantitatively different. Even leaving aside the assurance of multiple new platforms, applications, and interfaces (and, later, IPv6), managing AAA in a "best effort" environment with status rather than event information and a refocusing on stochastically modeled nongranular data is going to be a wrench. There is great complexity in managing multisession control and multiplatform/channel integration securely in an open network environment with unique customer identifiers. All of this needs to support a "single view of the customer" and "one instance of me" for all customers, where customer behaviors are enacted within multiple personal and affinity groups.

■ *Providing RA support for multidimensional convergence and continuous business transformation:* There is as yet no consolidated RA methodology for business transformation, nor have CSPs paid much attention to evolving skills require-ments, either during the interim/transitional phases or in the new generation environment itself. RA has a clear coordinating role as a business enabler—managing cross-functional semantics and methodology integration, as well as ensuring that opportunities for knowledge transfer, skills, and asset reuse, are exploited. One further note: business transformation needs to be man-aged on a "program" rather than a "project" basis. Its successful resolution will unavoidably be asynchronous with the concomitant underlying indi-vidual platform and systems evolution projects, which will have their own specific technical and business targets to meet. RA will need to safeguard this cultural and mindset evolution. This is because business leadership may tend to control the scope of projects as tightly as possible and to cut corners wherever business-wide policy and process changes are required for successful deployment.

... And all of this is monetized and is thus of great concern to all steadfast RA personnel. Finally, industry restructuring—with a move from vertical stovepipes to horizontal layers and all that this will entail—is lurking to a greater or lesser extent within the peripheral vision of many CSPs. Although the detail is outside the scope of this review, restructuring will lead to changing systems ownerships, departmental relationships, and inter-business processes, all of which will have the opportunity to create new revenue discontinuities that RA will have to fix.

* * *

Given the shift back toward an "accountancy-led" (as opposed to entrepreneurial) business leadership style, capex has retreated into limbo, and the opex focus is firmly on cost reduction. In reality, the hunger for resystemization and rationalization projects has not stopped and continues to grow, and the contribution that RA could and should make to the success of these initiatives is without question.

If CSPs are to compete—not just with each other, but also with the wide range of new organizations that convergence is bringing into the communications value chain, then the watchword of the foreseeable future needs to be "simplification." The principle of simplification needs to be applied not just to a reduction in the number and rationalization of the operational systems, but to pricing and tariffing, product catalogs, and customer interaction. Current CSP internal operating environments are just too complex to manage in any meaningful and profitable way in the longer term. The relationships that CSPs have with both customers and suppliers are also too complex. If simplification can be delivered as it should, it will be great news, not just for RA but for the ICT industry as a whole.

In summary, RA is and should always remain RA, but this should not limit the activities and influence of RA professionals within the business and operational domains. Under the guise of revenue management, and through the evolution of new generation ecosystems, RA is getting into new areas of activity and influence. Challenges in securing future investment represent a huge opportunity for both departmental and personal RA growth and development. At the same time, new skills and knowledge will be required, and there is a significant political dimension to be monitored, managed, and overcome.

What Makes for a Successful Revenue Assurance Team?

Mike Willett

In revenue assurance (RA) today, there is much discussion about the techniques for finding revenue leakage—the tools to use, where to look, how to use them. But in thinking about the factors that determine success, what are the attributes that differentiate the high-performing RA teams or individuals?

The answer will draw heavily on how you approach your RA work. I have a heavy bias toward a data analysis–orientated approach (in comparison to a process improvement approach), so let me call that bias out. Let me also point out, though, that a team based solely on data analytics expertise is not likely to succeed without other supporting areas. I want, however, to focus not on the skills and experience but on the innate talents or the things you cannot teach.

For me, then, there are a few essential attributes.

Resilience—I find this is key throughout the entire process from trying to acquire the necessary data and business rules to do the work, working through the analysis itself, understanding what real leakage is, and what noise is, and then advocating for any leakages to be appropriately addressed.

Integrity—Data may not lie, but if the RA team starts to make assumptions, does not question its own approach, and does not question the data it is getting, then the RA team can very easily start presenting false-positive results. When RA speaks, you want people to listen, not switch off.

Flexibility—Time and again what you set out to do and how you want to do it can easily be deemed impractical once the data starts arriving (or does not start arriving, as the case may be). To be able to answer the frustratingly simple question "Am I losing money here?" requires a clear focus and ability to adapt.

When you have a team with these attributes and the right strategic direction and support, then maybe, to steal from Eric, the caterpillar might well become a butterfly.

Dollars and Cents, Pounds and Pence

Eric Priezkalns

Let me share a little secret with you. As an RA consultant, I kept a little checklist of possible leakages to look for at any new client. My personal list has since been superseded by the work of people like Stephen Tebbett and Geoff Ibbett, who were the driving forces behind the TM Forum's leakage inventory, which is now an appendix to the TM Forum's *Revenue Assurance Guidebook* (Priezkalns 2009). Though my old list is now out of date, I still think it is worth sharing what I put at the very top. The first thing I looked for at a new client was evidence of a rating error where somebody had mistaken dollars for cents, pounds for pence, or any other similar confusion between units of currency. Most of the time I would be unlucky, but sometimes I would hit the jackpot and catch a business where somebody had misread a tariff change, or where the Pricing team had made a mistake when communicating it, or where the people managing reference data lacked sufficient training and did not feel obliged to double-check what they were doing.

When I got lucky and found an error of this type, then usually the error would be linked to an unusual type of call, like dialing up somebody's satellite phone. As a consequence, the revenues involved may be small and the total benefit limited. Even so, this particular quick win was very useful to me as a consultant. More than anything else, it is very effective at illustrating misperceptions about RA and hence why people should take it seriously. First, it would quickly alert any skeptics in the communication provider's management team to just how easy it is to make mistakes. These mistakes might impact the bottom line—or cause customers to get irate if they discover the error first. If I could convince management that a very simple error could go undetected simply because the impact was relatively small, it would make it easier for me to persuade them that complicated errors can go undetected despite doing much more harm. Second, the error with currency units shows that common sense is not enough. You actually have to look for errors rather than assume they are not there. An error of a factor of one hundred looks startlingly obvious when observed by someone with the right intuition for how much calls should cost, but the error will not be noticed if nobody thinks to look for it. Third, and most importantly, the issue vividly demonstrates how extremely simple *preventative* controls—like following and imposing simple conventions for how to communicate tariff changes—can save an awful lot of cost and bother later on. Detecting errors like this is a chore, so finding an error like this makes management think more seriously about the need to build integrity into their processes.

Time and attention put on simple, almost *boring* steps to stop mistakes are much better for the business than spending money on someone like me—or on software—to look for errors later on. There is a misconception that preventative controls need to be a burden. Not so. There is very little ongoing work involved in designing a pricing template that makes it clear what currency units were used. Do it right once, and you reap the benefit in avoiding mistakes forever more. Looking for errors after they occur is good for consultants and good for software vendors, but usually the best prevention techniques come from employees understanding what they do and then applying that to the care they take to ensure they do things right first time. These employees need not be dedicated RA staff; more important than the job title or the dedicated role is that people have the right attitude. Knowing the detail is more vital than anything else. If communication providers do not look to prevent the fault, they may not look to find it after it occurs. And if the communication provider's employees cannot spot the fault, there is no guarantee that software, or a clever consultant like me, will know to look for it either.

Never Say "No New Leakage"

Mike Willett

RA practitioners have many options to extend their analysis and assist their companies. If we are looking for "leakage" (and you will have to accept my definition) we set about identifying where business rules have not been properly implemented, resulting in an undercharge to the customer and subsequent lost revenue. This could include all manner of things—call records being lost, rating errors, discounts incorrectly applied, provision of services in the network but not in the billing system, and so on. Many of us continue to have enough work validating just these areas. However, do we know the marginal return from working in these areas as opposed to spending our time and resource elsewhere?

If you return to my simple definition of leakage, then this can be applied to other areas of the business beyond the traditional call/event flow, billing, and provisioning processes. What other business processes may be leading to leakage? At the top of my mind is the giving of credit adjustments to customers and understanding to what extent these are appropriate or otherwise. Do not think solely about fraud here either, but about whether the customer was entitled to a credit and also whether the amount given was appropriate. Another potential area surrounds the charging of fees—if there is need for manual intervention then there is a risk that this will not be fully charged. This mistake could relate to any fee, for anything that is charged a fee for, and so would be unique to the rules of each service provider. These are but two examples and I am sure you would all know of more.

If I can return to the leakage definition—it is focused on business rules being correctly implemented. RA should look at these rules to see if they make sense in the first place. The classic example is the hyper short-duration calls that came out of the fraud world. Numerous short calls are made to premium numbers—the duration is below a threshold for retail billing and so is discarded but is of sufficient duration for the premium service provider to expect payment. I am sure you can guess the rest. In this case, the business rule of discarding these short-duration retail calls is not beneficial to the service provider and rules should be changed (either in the retail billing or content settlement system) to close this gap. But there may be many other business rules that exist throughout processes and systems that were, in all likelihood, defined years ago for a now unknown reason. These rules may not just be suboptimal, they may be costly. Find some of these and make the case for a business rule change. Then RA is not just ensuring all entitled revenue is being billed and charged for—it is also growing revenue.

How Democracy Stops Leakage

Eric Priezkalns

I like the idea of democracy. I think there is something to be said for sharing authority between a large group of people rather than giving it to an elite or to a single ruler. That is why I like blogs and why I like the *talkRA* project. Why listen to one person's view when you can listen to many? There is wisdom in crowds, after all.

The problem with democracy is that if you give responsibility to many, you may find that, in the final reckoning, nobody takes any responsibility. Revenue assurance (RA) is like a public good. For a start, RA is something that benefits everybody in the communication provider (unless they actually want to work for an unprofitable and wasteful business that makes lots of mistakes). However, measuring and allocating responsibility for RA is virtually impossible to do. Many RA big shots would no doubt choke, or laugh, after reading that last sentence. But that is because they like to take all the credit for any successes, and take no responsibility for any failures. Holding an RA department responsible for RA is like holding environmental activists responsible for global warming, making the police responsible for crime, or holding a doctor responsible for the health of the community. Just because your vocation is focused on dealing with a certain kind of issue, it does not mean that ultimately you have more influence over it than the wider community does. The same is true for RA, and in this case the wider community is every employee of the communication provider, and quite often its suppliers too.

Of course RA practitioners have a vital role to play in stopping leakage. RA departments do not cause the problems they deal with (at least, we hope not). It follows that if you instill a culture where care is taken and fewer leakages are created, then there will be fewer leakages to worry about later on, and less work for the RA department to do. What is more, many of the solutions to leakages will fall outside of the control of the RA function. So if performance gets better, and leakage falls, then while the RA department may deserve some credit, much of the credit may also belong with other functions in the business. To continue the analogy, think about how stopping energy waste reduces global warming, or how taking better security precautions can prevent crime, or how regular exercise improves health. Like any public good, RA works best with the support and active engagement of the whole community for its own enlightened benefit.

Spending money on an RA department so it can employ people or so it can implement tools is like putting your faith in a centralized solution to a distributed problem. Leakage is a distributed problem, caused because distributed people make distributed mistakes. I refer to people because systems do not make mistakes. Only the people who choose systems, design systems, implement systems, and use systems can make mistakes; the systems themselves work the way they were made to work. There are lots of ways to avoid mistakes, like emphasizing simplicity in design, taking a modular approach, or being thorough with testing. Encouraging people to avoid mistakes or to identify and correct mistakes is like encouraging them to recycle or to exercise. It would be a bad doctor who wants his patients to be unhealthy just so that he can then give them expensive treatments, but it is very hard to measure the benefits of preventative medicine, and hence to reward it, in terms of either the doctor's efforts or the community's as a whole. Hard though it is, a communication provider must remember the importance of collective action, or else he will end up with very expensive doctors trying to cure a sickly business that never gets well. The conclusion is that RA is not the problem of the RA department—it is everyone's problem. This means the prime recurring challenge with RA is the same as with any public good: educate the community. If people do not know what the problem is, they will never start to work on the solution. If you get the community to address the problem, then things will get better. It is possible for everyone to share responsibility, and thus, to act responsibly.

Recognizing the Value of Revenue

Eric Priezkalns

It is an oddity of "revenue" assurance that so many practitioners have little or no knowledge of what the word "revenue" means. Paradoxically, if they are working for a CFO, they will be working for someone with a very thorough understanding of the word. Revenue is not just a simple function of sell more, get more revenue, company makes more profit, gets more cash, and everybody is happier. In fact, I have sometimes seen revenue assurance (RA) people embarrass themselves by their ignorance in front of their CFOs. Worse still, RA people sometimes do not even realize they are making fools of themselves. Ironically, these same people often talk about the need for support and sponsorship from the top, and will even give sermons about the importance of their jobs and the value they add to the businesses. Here is a newsflash for these people: if you want the CFO to understand your job, it is a good idea that you first show some understanding of the CFO's job.

Imagine the following scenario: the CFO is reviewing, yet again, the results for the year so far. He looks at the actuals, and the forecasts, and tries to know the reason for every variance. It looks like results are on target. Then the Head of RA strolls into his office for a meeting to explain what benefit his team has added over the year. There was no target for this Head of RA, but the Head of RA goes on to explain how his team added a few hundred million of revenue to the company's bottom line. Should we really be that surprised if the CFO's response is to be skeptical? Everything the CFO was measuring was on target. He is looking closely at the results on a regular basis. Everybody's numbers, without exception, come back to him. They all add up and he has explanations for everything. He has had forecasts, reforecasts, and more reforecasts. He has compared actual to budget, and this year to last year, and had every variance thoroughly explained. Yet, as if by magic, a team in his own directorate pops up and claims to be adding huge benefits that never previously were featured in any of his calculations.

Then imagine the scenario continues as follows: the CFO is impressed and is glad to see this level of return from his RA team, so he wants to ensure RA is managed like everything else—by forecasting the returns that the RA team will deliver in the next financial period. All of sudden, the Head of RA wishes he or she had never spoken up. To align expectations with reality, the Head of RA is confronted by one of two choices:

■ RA focuses on work that is predictable and easy to forecast. Work becomes dull "handle-turning," dealing repetitively with the same symptoms of the same known problems, but never fixing the root causes.

■ RA has to make wild guesses on the value it will add, without knowing what the leakages are, how much they are worth, or if the RA team will be able to prompt the necessary recoveries and fixes. In addition, this work must be done in a way so that the benefits can be shown to be genuine when they are scrutinized by the CFO.

Either way, from now on the CFO expects to see the numbers, and will expect that they are incremental to all the work performed elsewhere. The CFO will be disappointed if the numbers come short. Self-indulgent "measures" of RA benefits created by the RA team are of no use to this CFO, who wants all numbers to be analyzed and supported in the same way, so he can see how they add up and are consistent with the numbers reported for the whole business. This is a jolt for the RA team. Previously they calculated the benefits they added but their numbers were never reviewed by anyone else and they were never factored in the big analysis of what was up and what was down compared to forecast. Presenting unreliable measures, often devised ad hoc to flatter the RA team, is not an uncommon vice amongst RA departments in immature communication providers. Temptation gets the better of people. However, the communication provider's RA maturity cannot increase without proper measures of what real benefits were added. Remember, spending a lot on software may give people jobs, but it will not help the business if nobody can determine if the RA people are actually adding any value. Like any team, RA's reports of its performance must be properly and impartially reviewed to prevent the RA team giving a biased view of its own success. Which brings us back where we begun—with not understanding the word "revenue" and hence being unprepared for scrutiny.

I question whether any RA department can deserve that name unless it employs at least one person with a good technical understanding of the principles of revenue recognition and a detailed knowledge of the revenue recognition policy adopted by their business. Otherwise, the job of RA is not about revenue at all. The job is about data. The difference between revenue and data about revenue is that being sure about revenue figures means somebody must complete the loop and associate RA's figures with the revenues reported for the whole business, explaining the impact of RA relative to everyone else. An RA team that does not understand revenue should be called a *data integrity team* because they may understand the data but not what it is ultimately used for. You can check data without understanding its purpose. The shortcoming is that if you only partly understand the purpose of data, the checks may not be optimized for the needs and risks faced by the business. To claim to assure revenues, you must first know what revenue is, and all the factors that will influence the revenue numbers. This is no mystery; it is a matter of policy. Decisions relating to these policies have as much influence on the results as the work of RA, and the policies are scrutinized not just by the CFO but also by shareholders and corporate analysts. This leaves an RA department with little excuse to be ignorant of the policies or how these policies relate to the results that the RA team generates. Even so, lots of RA teams contain not a single person with more than a cursory understanding of the complexities of revenue recognition.

Assuring revenues means understanding revenues, and revenue recognition is a complicated discipline in its own right. RA people would do well to remember that.

If your RA team is not going to assure revenues, then it is best to not make claims that the CFO will see through. Instead, RA should be honest and explain the limits of what it does, including the limits on its own methods for evaluating the benefits it adds. When it comes to putting a value to the benefits added, RA departments take a chance when calculating their own numbers if they do not also understand revenue recognition. Some get lucky; others do not. To avoid the danger of having those numbers ripped apart at a senior level, if the RA team lacks the skills in-house, they should work with somebody in the business who really understands revenue recognition. Then, when the numbers are presented to the executives, those numbers will be more robust. They may also be smaller, but if you want executive support it is better to have a smaller and reliable number than a big number that the CFO should not trust.

How Big Should a Revenue Assurance Department Be?

Eric Priezkalns

Each axis on Figure 2.1 describes two key factors that influence the optimal head-count of a revenue assurance (RA) department.

One of the axes is automation. Automation can reduce the need for manual intervention in performing RA checks, and hence lower the manpower requirements.

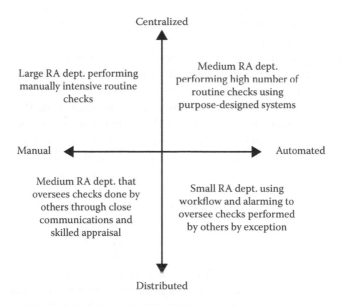

Figure 2.1   Revenue assurance department size.

In terms of maturity, it would be expected that automation increases over time, reducing the manpower burden in the long run. However, during early stages of maturity, automation and manpower are likely to grow hand in hand as the coverage of controls increases, leading to new tools and additional heads to execute the tests and reconciliations using those tools.

The other axis on this diagram is the degree to which RA responsibility is centralized. In some communication providers, the responsibility for performing detailed checks is completely centralized within a single RA team. In others, the responsibility for performing detailed checks is distributed around other operational areas, and the RA department acts as an overseer or supervisor, ensuring that others perform checks but not doing any checks themselves. An RA department that oversees a distributed environment only becomes directly involved if anomalies are reported to assist root cause analysis and to drive the resolution of problems.

Both the centralized and distributed approaches are perfectly valid, though the best choice will depend on the culture and organizational style of the communication provider. Irrespective of whether the tendency is to centralize or distribute responsibility for RA, the maturity model does suggest that the necessary overall manpower will vary with maturity. The total manpower applied to RA will grow in its earlier stages, reach a peak as near-complete coverage is reached for the first time, and then decline as there are diminishing returns and activities reach an optimal level of efficiency. This assumes that the RA scope remains unchanged in the meantime.

The maturity model also discusses the interactions between the central team and the distributed agents performing RA tasks. In general, it is assumed that the central team will become more active and prominent as new and comprehensive detection-based approaches are implemented. This reaches a peak at the mid-level of maturity, level 3. However, responsibilities and use of detection tools may be increasingly distributed as the business reaches level 5.

Whether the business is homogenous or heterogeneous is another factor that determines the necessary head count and will also alter views on maturity. A homogenous communication provider will exhibit less variety in general; it has fewer products, more consistent network technology, a smaller number of BSS platforms, and so on. A heterogeneous communication provider is defined by having some major distinctions. The most straightforward example is a multiplay provider where both providing and charging distinct services like fixed-line, mobile, Internet, and television are really quite separate, even if presented on a common bill. Similar traffic may be handled by a number of different legacy platforms. Another way of talking about homogeneity or heterogeneity is in terms of complexity. A more complex business is more heterogeneous, while a simple business is homogeneous in nature. Whatever terminology is used, the main question to ask is whether the business is best described as one consistent series of operations or whether it is better described as a series of loosely interrelated but distinct operations.

If the communication provider is quite heterogeneous, this will lead to a multiplication of the people, and possibly also of the systems, needed to perform RA for each part of the business. In such cases, it may be necessary to treat the communication provider as a "constellation" of different businesses, each with its own distinct products, technologies, markets, and risks. For example, there will be similarities but also very significant technical differences between assuring a cable television offering and assuring the mobile voice offering of a multiplay provider that offers both. Other reasons to segment a business might be because of different technical architectures and different pricing and marketing cultures for prepaid and postpaid products, or assurance may need to be segmented to reflect a variety of legacy systems, or because of the distinct kinds of networks being operated. Geography may also be a factor that creates heterogeneity. This is most obvious for an RA team that works at group level in a multinational group. Operating companies will exhibit different cultures in different countries and will very possibly sell unrelated products using their own distinct OSS and BSS technologies. One final factor that causes heterogeneity is merger and acquisition activity. Where a communication provider is born of the coming together of two separate businesses, each with its own operational history, the assurance task cannot be synergized until other synergies are realized in how the operations work.

When faced with heterogeneity, an RA department may need to deliberately organize itself into separate teams that map to the separate divisions of the communication provider. This is most appropriate where the RA department works in a provider that is most like a "constellation." The result is that the RA department is also likely to be a constellation with parts of the team specializing in quite different products and systems. Their techniques may vary greatly. If this is the case, it may be appropriate to separately set targets, separately measure performance, and separately assess the RA maturity for each different business line. Of course, few real-world providers will perfectly correspond to either a heterogeneous or a homogeneous model, and it is up to the RA practitioners to both find what is common while also having the versatility and adaptability to deal with the complexity they face.

Chapter 3

Linking Revenue Assurance to Fraud Management

Synergy of Billing Verification and Fraud Management

David Leshem

In this article, I would like to provide a different perspective on the supposedly separate disciplines of fraud management and billing verification and to show that there is significant synergy to be gained by an integrated approach.

Fraud management systems (FMSs) and billing verification (BV) systems are among the technologies available to revenue assurance (RA) managers at communication providers. Each technology tackles a different aspect of RA. Fraud management handles mainly external attacks on the communication provider's revenue, while BV handles internal failures due to human error, misspelled product/tariff specifications, or misinterpreted business rules.

Fraud management is a "cat and mouse" game between fraud perpetrators, who are usually well-organized and technologically sophisticated criminal groups, and the telecom service providers they seek to defraud. Fraud management comprises a large family of techniques that aim to tackle the wide variety of types of fraud.

However, there is no final victory possible over fraud. This is true for two reasons:

(i) New types of fraud are regularly invented, and new fraud opportunities are created by the service and technology advancements of communication networks.
(ii) Finite technological and manpower resources set an upper limit on the capabilities of fraud management to either prevent or detect fraud.

Fraudsters are inventive and are always exploring new techniques to exploit the weaknesses of communication providers. As communication providers develop new products and implement new systems, the fraudsters search for new loopholes to exploit. At the same time, FMSs only discover new types of fraud after the damage has been done. They are expensive to purchase. They are also expensive to operate, as the user needs to be highly skilled to get the most from the FMS. One major shortcoming of FMSs is that they cannot accurately prioritize on the basis of a quantified valuation of the fraud threat. As a consequence, fraud management teams may waste much time chasing the less costly frauds.

RA and BV systems have traditionally been viewed as financial audit tools. They have not been typically been used to supplement fraud management. I would like to show that BV systems could enhance the capabilities of existing FMSs in the following ways:

(a) Detecting frauds missed by the FMS
(b) Detecting suspected fraud patterns earlier
(c) Helping tune the FMS's detection algorithms in relation to changes in legitimate customer behavior to avoid false alarms
(d) Assisting with content and e-commerce fraud prevention

The BV system's ability to monitor and reconcile all of the communication provider's revenue streams is important. Also important is the BV system's prioritization of the sources of revenue loss. Together, they enable the communication provider's fraud management experts to focus their efforts in fruitful directions.

Both FMS and BV systems process CDRs at a different stage and format. FMS process mainly raw CDRs or SS7 events, while BV systems process CDRs that are revenue-bearing events and contain customer-related information for billing (e.g., calling circle discount indicators, cell information, package information, and much more). After the initial step of CDR collection, the information is processed quite differently in these systems. In addition, technological experts usually operate FMS, while personnel with a financial background operate the BV systems.

I want to think that the performance of FMS can be improved by effectively using the capabilities and information available from BV systems.

The main desired features of a FMS are the following:

- Real-time monitoring of the network to enable proactive prevention of fraud as it happens, rather than reactive "management" of the fraud
- Coverage of a maximum number of the communication provider's customers and of different types of fraud
- A minimum of false-positive alarms
- The ability to detect new types of fraud quickly
- Efficient case management
- A minimum total cost of system ownership (TCO)

BV systems can play two major roles in the fraud management process by

(i) Offering enhancement of the fraud detection capabilities of the FMS
(ii) Interfacing the FMS to the business processes of the communication provider to achieve a better overall systems integration

The BV system has the capability of detecting unbalanced revenue that flows into and out of the communication provider's systems. Such an imbalance may result from a number of causes, one being fraud. So the basic question that needs to be asked is whether there are fraud types that create such revenue imbalances. The answer is positive. The following are two examples: one of fraud and one of service abuse.

Example 1: Internal Fraud

Internal fraud is a generic name for several subtypes of fraud. These frauds are tough to detect because they are perpetrated by people from inside the organization who know where the network's loopholes are and because the frauds are usually not detectable by systems that rely on obtaining CDRs from SS7 signaling probes. The examples illustrate some typical internal frauds and how they affect the revenue flow:

- An account is provisioned in a network switch, but not in the billing system, meaning that calls for the account are never billed.
- Incorrect setting of subscriber service category means calls are charged at a lower rate than they should be.
- The affected communication provider has costs to pay to various interconnect and content partners in relation to services provided, yet does not collect the associated amounts due from the retail customer.

Example 2: Service Abuse

A communication provider does not charge the retail customers for the first few seconds of use of a particular premium rate service. This enables fraudsters to install automatic call generators that place a high volume of short duration calls to accomplice premium rate service providers in another country. The communication provider does not collect any retail revenues relating to the fraud but has to pay for the calls to its interconnect partners. This is a classic premium rate fraud, but in this case the fraudsters were simply taking advantage of a service loophole and were not breaking any laws.

Conclusion

These types of frauds and service abuses would be discovered by the BV system within one billing cycle. In real-time systems they would be discovered much earlier.

In some cases it may be possible to replace FMS algorithms with BV system rules and rate parameters, reducing the processing load on the FMS. For example, one of the toughest and most prevalent types of fraud is subscription fraud, where a false identity is used to open an account with a communication provider and to obtain services until the communication provider realizes that there is no intention to pay for such services. A BV system could check the reasonableness of the call pattern by analyzing the following:

- How fast are charges accumulating?
- Are charges accumulating at all times of day and night?
- Are calls all to the same destinations?
- Is the use of particular services unusually frequent?
- Does the calling pattern match known patterns of behavior of legitimate customers?
- Does the calling pattern match known patterns of fraudulent behavior?

This check, based on the single variable of price/revenue, would be easier than a multifield check of CDRs.

When Fraud Belongs with Revenue Assurance

Eric Priezkalns

There is a debate about whether fraud management falls within the remit of revenue assurance (RA), or has anything to do with RA, or is something that needs to be managed differently. I never could understand those arguments. For the most part, those arguments are too theoretical, or they are based on too limited a series of practical experiences. The golden rule is always that businesses should organize themselves in the way most efficient for their circumstances. Sometimes that will mean linking the management of fraud with RA; sometimes it will mean keeping them separate. It all rather depends on what frauds and revenue losses the business is likely to suffer from and how it deals with them.

There is a world of difference between an inadvertent but costly flaw and the deliberate exploitation, or creation, of a weakness by someone with a financial motive. Many leakages revolve around similar weaknesses or gaps in monitoring, and there is overlap in the indicators of fraud and revenue loss. Nevertheless, fraud threats are often ignored—perhaps in the hope that somebody else in the business will deal with them. The upshot is that some frauds are ignored completely. Fraud management has linkages to RA but it also has linkages to security and risk management. This begs the question of whether greater synergy is achieved by locating the fraud management team with the team responsible for security.

Of course, in the real world lots of questions about who does what job are decided by internal politics and power struggles. But if you want to think about the question solely on the basis of efficiency and effectiveness, here are the three factors to consider when deciding whether to link the management of fraud with that of RA.

(1) *Human Resources, Staff Time, and Communications.* Here are some questions to ask to help judge if fraud management could be performed alongside RA:

■ Will there be prelaunch analysis of risks for both fraud and RA?
■ If yes, could this be performed by the same individuals?
■ Can the training of fraud and RA analysts be combined to improve effectiveness and reduce overheads?
■ If fraud and RA staff were in the same function, would they have improved promotion prospects, making it easier to retain good people?
■ Are losses due to fraud and revenue leaks reported using a consistent format?
■ Are the losses due to both presented to the same executives?
■ Could loss reporting be collated and reported by the same person?
■ Are losses calculated in a consistent way?

If the answers to the questions are "yes," linking fraud and RA will provide efficiencies in terms of human resources, communications, and the time of senior staff. If the answers are "no," it is still worth asking if there are good reasons for that answer. Why not have the same person trained to identify fraud and RA risks? Why not calculate losses in a consistent manner and present them to the same executives?

(2) *Incident Management and Process Improvement.* Here are some further questions to ask:

■ Are there overlaps in the way fraud and RA weaknesses are addressed?
■ Is there a similar scope for the systems and processes reviewed and monitored for both fraud and RA purposes?

- Is the same or similar documentation used to understand internal processes and system performance and hence to identify weaknesses?
- Are there topics like information security or business continuity where both fraud and RA considerations are often addressed in the same way?
- Do improvements related to reducing fraud often have benefits for reducing revenue loss and vice versa?

If the answers are "yes," there is a good case for trying to resolve issues and fix bad processes using a common approach and common prioritization. Where the understanding of technology and products is much the same for preventing or identifying weaknesses relating to both fraud and revenue leakage, duplication of effort can be reduced by linking the goals.

(3) Monitoring Systems and Processes.

- Is the same source data used for both fraud and RA monitoring?
- Are alarms that indicate possible fraud sometimes set off by accidental revenue leakage?
- Are checks for accidental revenue leakage sometimes evidence of deliberate fraud?

If the answers are "yes," there should be cost efficiencies in implementing systems and processes able to identify and react to both deliberate frauds and accidental revenue losses. Exploitative customer actions such as use of SIM boxes, or internal frauds where some customer bills are suppressed, may not be picked up by classic monitoring focused on looking for certain predefined types of customer fraud, but may get identified through generalized RA checks. Likewise, monitoring for unusual customer activity or for internal frauds may highlight accidental errors by the business or one of its partners, as well as identify genuine frauds. If there is a high degree of overlap in the recurring checks that would counter both fraud and revenue loss, then it is more efficient to combine the monitoring strategies and reuse data where possible.

It is worth noting that fraud management often addresses a much broader scope of threats than RA. This may be disguised in practice because responsibility for different kinds of frauds may be given to different kinds of teams. A security function will tend to carry the brunt of responsibility for identifying internal frauds, while a fraud management team in a customer-facing function such as customer services will inevitably have more of an external orientation to the frauds it considers within scope. Linking fraud management to RA may, if handled poorly, set a limit on the scope of fraud management that leaves gaps in the coverage of fraud management. RA functions typically have a history of overcoming skepticism. It is ironic that an RA function may unintentionally set inappropriate limits on the company's strategy to mitigating the risk of fraud. It is possible to find synergies in the management of fraud and RA, but even where synergies are worth pursuing, it must not be at the expense of leaving gaps that fraudsters will exploit.

Greatest Internal Fraud Risk?

Eric Priezkalns

The following checklist gives warning indicators relating to opportunistic internal fraud:

- The internal control environment is weak.
- There is regular turnover of key staff.
- Policies are weak, poorly documented, and rarely followed in practice.
- Procedures are not well defined and staff are not trained or expected to follow those procedures in practice.
- Staff has inside knowledge of how processes work.
- Individual members of staff occupy roles where there is a high degree of trust.
- There are strong ties between staff and the employees of suppliers.
- Work is rarely routine; staff often tend to be responding to one or other crisis situation.
- Staff do not enjoy good interpersonal relationships with each other and management.
- Staff suffer from low morale.

Take a look at that list once more. A department that suffers from most of these problems would be a fertile territory for internal fraud. Now think about the typical revenue assurance (RA) department. Do they know about the processes? Yup. Are they in a position of trust, with access to sensitive data and the right to direct alterations to it? Yup. While they give out controls to others, might they lack adequate controls, especially over the nonstandard activities and irregular projects they conduct? Yup. Do they have weak policies and procedures, especially when taking on new challenges? Yup. Can retaining key RA staff sometimes be a problem? Yup. Does RA work as if the sky is falling in? All too often, yup. And low morale? Sadly, the answer to that is often "yes" as well.

Then think some more about the internal fraud threat posed by the RA department. Staff working in RA have some very particular advantages if they were inclined to engage in fraud. They may be free to demand and make changes to network and billing data, especially during data cleanses and system migrations, without any proper supervision or review to ensure those changes are correct and justified. RA departments may have the liberty to set up and remove services and accounts. They may have access to SS7 probe data and be able to track the calls made and received on specific lines. They may even be able to snoop on private messages like SMS texts. They will often be responsible for test accounts that are an exception to normal billing activities and hence may not be as well controlled. Many of the staff will have superior skills for manipulating data or a keen understanding of how changes to data may exploit controls weaknesses to the benefit of a customer, who they may be colluding with. Because of recruitment pressures and because the numbers of staff are relatively low, checks on staff may not be as stringent as in other areas more commonly perceived to be high risk. In addition, vendor, contract, and consultant staff may be given unusual levels of freedom to inspect and even alter sensitive data. And finally, the line management responsibility for fraud management and RA may be unified. A failure to segregate duties magnifies the risk of opportunistic internal fraud executed under the guise of RA.

Of course, I am not saying that all RA staff are fraudsters, but they have an unusual freedom and latitude to access systems and data, and to make changes, and with that comes the danger of fraud. So the question is, really, what are fraud management departments doing to mitigate the particular and unique risks relevant to the RA department? If they just trust that RA people care about the bottom line, and cannot be corrupted, then they are missing the point. You cannot simply rely upon staff to be good just because their job requires it. When fraud and RA report to the same manager there may be nobody else within the organization expert enough, and with sufficient authority, to ensure that there is no fraudulent collusion between the fraud management and RA functions. All of these points raise tricky and uncomfortable questions, but staff who work in fraud prevention or RA should aim to provide good answers.

Chapter 4

Strategy, Scope, and Synergy

Five Dimensions of Revenue Assurance

Mark Yelland

The concept of describing revenue assurance (RA) in a five-dimensional model was first presented in *Revenue Assurance for Service Providers* (Yelland and Sherick 2009). This article looks to build on those ideas by providing a little more explanation behind the thinking.

The five-dimensional model was designed to provide communication service providers (CSPs) with a starting point for both usage and nonusage RA and to assist those with an existing approach to develop a coherent and structured enhancement of their RA capabilities, ultimately providing better integration within the business.

Briefly, the five dimensions are as follows:

1. Completeness and accuracy of records
2. Completeness and accuracy of charges, both incoming and outgoing
3. Identification of potential issues within margins
4. Improvements in cash flow
5. Creation of a business differentiator through customer-specific revenue assurance

Each of these is described in more detail subsequently.

Dimension 1: Completeness and Accuracy of Records

The single dimension is simply one where progression is along the revenue generation chain for a product or service. It takes the input to one process and ensures that there is a corresponding output, or number of outputs. It takes these outputs and ensures that each output is accurately reflected as an input to the next stage. The process is linear in that the only information required is provided by the processes, there is no requirement to combine information from two different sources to arrive at an output. The result of a one-dimensional analysis is confidence in the accuracy and completeness of the information transfer process.

For example, consider a typical broadband provider as illustrated in Figure 4.1. By taking the orders from the customer order system, one can track the orders into the provisioning system, through the network installation and ensure that all the orders appear in the billing system.

In summary, the first dimension of RA involves reconciling the flow of data from a customer instigating a chargeable service or order to the customer being charged for it.

Dimension 2: Completeness and Accuracy of Charges, both Incoming and Outgoing

The two-dimensional approach takes information from two discrete and unrelated sources—for example, the service and the tariff. It combines the information to calculate what the monetary charge should be. This is then compared with the actual charges shown on the appropriate invoice. A single service or product may incur multiple charges and outpayments.

For example, see Figure 4.2 and consider the different tariffs applied to a retail and business consumer for a similar international call.

To check the accuracy of the charge it is not adequate simply to know the details of the call. A second piece of information, the tariff schedule appropriate to that specific customer, is required. This cannot be obtained from the usage data.

Note: If we assume that systems are generally reliable, that tariff A times service B will always generate answer X, then there is no requirement to routinely check

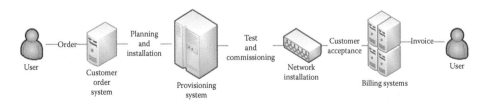

Figure 4.1   One-dimensional revenue assurance.

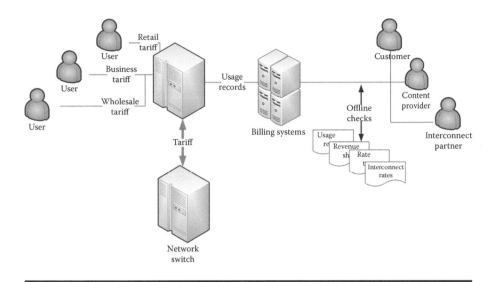

Figure 4.2   Two-dimensional revenue assurance.

100% of the services and 100% of the tariffs. The alternative assumption is that systems are unreliable and generate effectively random results when combining tariff and services, in which case you really do have problems!

In summary, the second dimension of RA involves confirming not only that the customers are charged for each and every service or order they get, but also that the charge is the right monetary value for that service. This means bringing together the data in the first dimension with additional data about the tariff that should be applied.

Dimension 3: Identification of Potential Issues within Margins

The third dimension introduces the concept of planned and unplanned leakages. Unplanned leakages are those inaccuracies that inadvertently occur in the day-to-day operation of a business. Planned leakage is money that the CSP chooses to give away through activities such as promotions, incentives, and discounts. This planned leakage is often not recognized by the business and hence not managed. From a business perspective, the marketing decisions are made on the basis of predicted customer behavior. In response, the RA activity should be able to provide independent assurance that the actual customer behavior exhibited by those taking advantage of the promotion has not invalidated the marketing assumptions.

For example: Evening calls to Australia for up to 1 hour cost a flat retail charge of £0.10. The preoffer behavior could be that there is no net interconnect outpayment between the United Kingdom and Australia; individuals spend as much time calling the United Kingdom from Australia as the individuals in United Kingdom spend calling Australia. There is traffic, but it is balanced. The promotion results in two changes, the first is that calls from the United Kingdom now last longer with many lasting the full hour, the second is that the number of calls originating from Australia drops and is replaced by additional calls outgoing from the United Kingdom. These two combine to turn a balanced flow of minutes into a significant net outflow of traffic from the United Kingdom. This results in a net outpayment to the far end by the interconnect team that may not have been considered by marketing when they designed this retail customer promotion.

There are other similar examples concerning bundles and free minutes, where the issue is not the accuracy of the billing, but the net impact on the profitability of the products themselves.

In summary, the third dimension of RA builds on the second dimension not just by ensuring that customers receive the right charge, but by also comparing these charges to the costs incurred in supplying a service.

Dimension 4: Improvements in Cash Flow

The fourth dimension recognizes that, in business, cash is king. With its cross-functional remit and access to data sources, the RA function is well placed to take ownership of activities to improve the cash flow.

For example: An analysis of billing queries would reveal two issues. The first is that there is an underlying series of problems that need to be resolved—this can be ascertained using conventional problem solving tools such as Ishikawa diagrams, root cause analysis, 80:20 Rule, Brainstorming, and so forth. The second is that payments are being delayed while the queries are being addressed. Therefore, even when a series of billing queries results in no credit notes being raised, the CSP's cash flow is being adversely impacted. Hence resolving the underlying causes should result in invoices being paid more quickly and a reduction in the debtor days for the business.

In summary, the fourth dimension extends the three-dimensional analysis by also evaluating the timing of actual cash flows.

Dimension 5: Creating a Business Differentiator through Customer-Specific Revenue Assurance

The fifth dimension considers the relationship between the CSP and its major customers. With many CSPs there are a few significant customers who generate most of the total traffic and revenues. The profitability of the CSP may be driven by the top 10 or 20 wholesale or corporate customers. For the CSP, the impact of such a customer deserting and moving to the competition can result in a significant impact on the bottom line. The traditional way this has been handled is to offer inducements such as lower rates as a means of encouraging them not to churn. The problem with this approach is that it negatively impacts margins. By developing a bespoke RA plan for these customers, it is possible for both parties to win, without necessarily compromising the margins involved.

The logic is that larger organizations employ people to check their telecoms invoice and raise queries if something looks amiss. This results in the customer spending his own money to check that the invoice is accurate. If one looks at the successful manufacturing organizations, it becomes clear that they have moved away from inspection on receipt, replacing it with an agreed set of tests/checks that are required of the supplier, along with a signed statement that these checks have been performed and the results are satisfactory. The benefit for the manufacturer is reduced internal overheads, the benefit for the supplier is that goods are accepted by the manufacturer more quickly, resulting in invoices being paid more quickly. This approach can be adopted within any supplier chain; it also works between a customer and its CSP. The benefit for the CSP is that until the competition adopts similar practices, the customer knows that moving to a competitor will result in increased costs and headcount within their own organization or increased risk of being overbilled.

In summary, the fifth dimension turns RA into a marketable service in its own right, offered to customers as a way of adding value to their business and reducing their costs.

In this article, I have tried to expand a little more on the five-dimensional model that David Sherick and I developed for RA. The dimensions represent discrete levels of performance, with increasing benefits arising as the number of dimensions increases. It, therefore, represents a potential path for aspiring RA teams just starting out or already some way down the path.

Exploring the Concept of Maturity in the Performance of Revenue Assurance

Eric Priezkalns

Revenue assurance maturity is one of the most quoted and discussed concepts in the industry, but it is also one of the most misunderstood. The goal behind a maturity model is to provide a framework for understanding RA that emphasizes the following three concepts:

1. The level of performance of RA is determined by the whole business, not by any function in isolation.
2. To improve RA performance overall, it is necessary to improve all the enablers of RA.
3. Improvement in RA is not attained in a straight line—each business needs to go through distinct learning stages and has to attain specific goals before it is ready to take on new goals and make progress toward the next level of maturity.

This article will go through the RA maturity model, its five levels and five dimensions, and the typical evolution path and obstacles to improvement. This will lead to some worked examples of how to use the model to appraise RA maturity in practice, using an all-new questionnaire designed to enable staff to perform a rapid self-appraisal of RA maturity. In turn, we will discuss how to use the results from the appraisal process to analyze and set the priorities for the business from an RA perspective.

The concept of using a five-level maturity scale is far from new. The principle predates RA but was adapted for RA purposes. To this author's knowledge, rudimentary maturity models were constructed by several communication providers (CPs) in the early noughties, most notably by T-Mobile, UK. The idea for a cross-industry collaboration to develop the model was discussed at the first TM Forum RA team meeting, which was held in 2004. At that meeting, it was the T-Mobile model that was subsequently donated and developed into the TM Forum's version of the maturity model. In parallel, other operators adapted and separately developed their own models, all with a similar structure but different detailed contents. In practice, several flaws have been found with the most common questionnaires in use, including that of the TM Forum. These difficulties include problems with adapting the questions to varied business models; multiple possible but inconsistent interpretations of some questions; some vagueness in wording; and a repeated tendency to answer the questions solely from the perspective of the nominal RA department, and not on behalf of the whole business. The new questionnaire presented here seeks to address those deficits.

Before examining the questionnaire itself, it helps to be walked through the conceptual framework and the mechanics of how a maturity model works. The goal is to deliver an analytical tool that is deceptively easy to use, but based on some very sophisticated analysis of what happens not just in CPs but in many businesses confronted with a challenge of how to improve their performance in the face of complex interactions between technology and people.

The aim of the article is to provide

- A grounding in how to use the model to assess RA maturity
- An overview of the main obstacles to increasing maturity (including the ones that people usually prefer not to speak about)
- A basis for understanding how to look beyond the nominal RA department and to instead treat RA as a holistic series of activities with implications across the business

We will also cover the ways to use the model as a tool for setting priorities, measuring progress, and even benchmarking compared to peers, all while retaining a longer-term strategic perspective on RA. The model can also be used as a communication tool, and this will be explained throughout. In addition, the detailed discussion of the model will cover the interplay between the different aspects of maturity within an organization and ways to ensure that goals are realistic and balanced.

Perhaps the most important reason for assessing maturity is that it is a way to reflect on the extent to which the mission of RA has been accomplished, what is done well and what is not being done. The appraiser is required to look at the business as a whole, and consider the extent to which the risk of revenue leakage has been mitigated all throughout the business. This of course supposes they know the attitude throughout the business; it is possible to draw useful inferences just from the fact of RA not being known or appreciated outside of a small caste within the CP.

The maturity assessment considers a series of parameters or enablers, all of which need to be in place to get the best results on a permanent basis. By leading the appraiser through all parameters, it ensures that consideration is given not just to the most obvious or immediate causes of concern, but to all factors that will influence the success of RA in the long term. One of the benefits of the exercise is that the appraiser is encouraged to see RA as more than just a reaction to specific and current issues, even if that is how RA is usually treated in their business. To break out of a reactive cycle of devoting all resources to the latest problems or haphazardly identifying and pursuing new objectives, the maturity model encourages a more considered approach to planning. It also assists in the realization of plans by giving a context to communicate those plans.

We need to be clear on a few things that the maturity model does and does not do. Over the years, I have become very wary to manage the expectations of people who want RA training. You get the feeling that some people, especially if they are new to RA, would like to hear in a few hours or read in a few words all of the most important answers to the question of how to do RA. This is not realistic for a couple of reasons. The first and most obvious is that there is too much to learn for a novice to master it all in a few hours, never mind the fact that whatever is learned needs to be interpreted and applied to the different specific challenges in any given CP. But it is just as vital to realize that the priorities of an RA function will change over time. When establishing a fledgling platform for RA in a CP, the goals are quite different to those which are relevant when RA has become established or when RA faces the challenge of diminishing returns. The maturity model recognizes this by setting nonlinear expectations of what is needed to go from one maturity level to the next.

In the early days of RA you really had to read around the subject to learn, because there were no resources about how to do RA. It is unrealistic that in the meantime somebody should have come along and written a one-stop-solution on how to do RA—if it was as easy as that, then everybody would have copied it, everybody would have done it, and everybody would have made big bags of money from following the instructions in that manual. Fortunately for us, there is a lot more to RA than that, or else nobody would have a very long career doing it. What the maturity model gives is a strategic framework that informs the major priorities for RA. Within this strategic framework, the RA practitioner has to identify and apply the right tactics and operations that best suit their CP's culture, systems, processes, needs, and challenges.

So what does the maturity model do? It explains a strategic evolutionary path that would deliver deeply embedded RA in the long run, by pointing out a series of steps starting from wherever you are now. It is not a "how to" or "self-help" guide for RA. What the model does, and does very well, is to place RA into a strategic context that can be applied to any CP's business, and provide a simple but powerful analytical tool to help CPs appraise their own maturity and derive strategic priorities accordingly.

The following are the essential basics of the model:

■ It has five levels and five dimensions.
■ The levels are the levels of maturity, from least mature to most mature. We count from 1 for the least mature to 5 for the most mature. Later on we will talk about scoring maturity, and for consistency, all the scores will be from 1 to 5, with 1 being least mature and 5 being most mature.
■ The model hence represents a series of steps from 1 to 5. The goal is to climb up the steps to get to maturity level 5. At each step, the maturity model describes the typical activities and behaviors of a business at that level of RA maturity. No business will be an exact match to the descriptions given. The idea is to compare the real CP with the descriptions, and hence to use them as a way to identify your current degree of maturity and how to step up to the next level.
■ There are five distinct aspects or categories of business operations, performance, and behavior that define the potential for successful RA.

We will come back to the five aspects in a little while. First, we shall discuss the maturity levels.

Maturity Levels

Level 1, the initial level, is the least mature. Level 5, the optimizing level, is the most mature. The intermediate levels are called "repeatable," "defined," and "managed." Those names might sound familiar—we will discuss that later.

Some could argue that we should include a level 0, for businesses that literally do no RA. That is fair, but we did not do so because if you are interested in the maturity model, chances are you already are doing a little RA. Anyone applying the model will either be at level 1 or have been through level 1. Level 1 is the starting point. At level 1, the key challenge is getting buy-in and building momentum. The main benefit of the maturity model is that it gives the practitioner some perspective on the long-run evolutionary path of RA in other CPs. Using the maturity model as a basis for comparison can help with securing buy-in for a more ambitious RA program than might otherwise be the case.

The top level of maturity—the optimizing level—is called that because it is the aspiration. One important decision about this model is that it should be very difficult for any CP to reach this level, and a struggle to remain at this level even if reached. This is in strong contrast to some alternate models that suggest an easy progression to level 5—and then beg the question of what needs to be done next! At present, it is not believed that any CP will have achieved level 5 as defined here. However, we know that some might be likely to claim they have; as often as not, such claims are based on a profound misunderstanding about what maturity is. The whole point of level 5 is that if you get to this level, you cannot get better, and you need to work hard to stay at this level. It is hence right that this is an aspiration that will be a struggle to be realized. The maturity model is meant to stay relevant even to CPs who have done a very good job in realizing RA throughout their operations. It is a strategic tool—if it is to be useful, it has to stay relevant even as CPs climb up the steps. The climb is mountainous. As the CP gets nearer to the top, it gets harder to keep going up, and easier to fall down. It is not a race. Once you finish a race, you have nowhere left to go. The challenge posed by this model is to reach the top of maturity and then to stay there.

At level 1, RA is ad hoc and chaotic. Some things get done, others are not done. Controls and checks may be performed one month but not the next or ever again. There is no clear logic or prioritization that determines what is done. Much is based on guesswork or on just what is the easiest thing to do with the data, tools, and people available. A lot depends on individuals who feel highly committed to doing RA. They may find they are balancing their desire to do RA with other demands on their time, because they can see the business has issues that need to be addressed.

At maturity level 2, the main difference is that RA becomes repeatable. Some things that should be done are still not done, but the activities—for example, reconciliations, tests, and trend analyses—that are undertaken are performed consistently and on a regular basis. If somebody leaves the company, somebody else will fill the gap and take on the RA tasks of the leaver, performing them in a consistent way. There may be no great vision for how to perform RA, but there is basic project and process management. There will be gaps in controls, some of which are recognized, and others go unrecognized. However, management, if imperfect, can be expected to give regular attention to their responsibilities to check and verify basics about the accuracy and integrity of revenues, costs, billing, and accounting. No effort has been put into reviewing how efficiently the tasks are performed, the importance of what has been done versus what is not done, or into mapping and communicating the interdependencies between the people doing RA and the people and systems they rely upon for data. There is also little work done on root cause analysis, and when problems are detected, it may be a struggle to determine what action should be taken and who should take it.

At level 3, the work and relationships between a dedicated RA team and some other key parts of the business are formalized. While we talk about a dedicated team, it is possible that the scope of RA is allocated to two or more dedicated teams in separate parts of the business, but a single team is most likely. Dedicated RA staff will have responsibility for a broad spread of detective controls, monitoring a good proportion of revenue streams, and highlighting where mistakes are occurring. Their work is still predominantly reactive, though the reaction times may be a lot lower as monitoring tasks may be performed daily and new alarms may work in near to real time. Gaps in control coverage are fewer, and the gaps are consciously recognized. The team reports values for the work it does, but for the most part these cannot be easily tallied to the figures in the management accounts, as it is very difficult to separate out the quantifiable benefits delivered by the investment in RA. The easiest activities to quantify are preferred: those that result in back bills or other correction of historic errors where it is simple to calculate the financial benefit by inference from the numbers of calls, customers, lines, and so forth, that were not being charged or were being charged in error. This may lead to a bias toward assuring postpaid revenue streams over prepaid streams. The dominant slant of RA is still to find faults after they occur—although controls are now more "active" rather than just "reactive," meaning for postpaid business models RA catches and resolves problems before the billing date.

With level 3 of maturity, the reactive style of RA reaches its pinnacle. Progression to level 4 and beyond depends on a significant change of tack, with emphasis shifting to prevention as the major driver of future improvement. This comes with a realization that more resources should be applied to root cause analysis, preempting problems before they occur, and the designing of RA controls to be implemented alongside new products and technology. An element of the resource dedicated to RA becomes more forward thinking, trying to reduce the inflow of errors detected by the reactive controls by permanently fixing root causes and by evangelizing the risks and causes of error, especially during any major product launches or transformation projects. The reliability of data and reporting from the reactive side of the team helps to increase awareness of the importance of RA.

The change from a detective to preventative approach to revenue leakage is an insurmountable challenge for some RA departments. This occurs not least because the expectations for staff skills and inclinations are quite different. Within the industry, there is quite some pressure to develop an alternative view of maturity that sees levels 4 and 5 of maturity as being marked by increased sophistication in detective controls, with relatively little emphasis on prevention. However, this contradicts the thinking behind the maturity models that were the prototypes for the RA maturity model. In addition, some RA functions do successfully transform their approach to embrace a much more preventative view of how to perform RA. In the final reckoning, a maturity model based solely on detective controls would only encourage RA functions to go down a strategic dead end.

At level 4, the reactive mode of RA is complete, with perfect coverage of detective controls across all possible leakage points in real time, and with layers of controls complementing and acting as back up to each other. The last additions to the reactive controls may no longer be cost effective because the cost of implementing the extra detective controls may be higher than the value of leakages they find. Because detective controls are complete at level 4, the quantitative value of leakage detected is well understood. Improved understanding of the relationship with accounting means RA performance can be tallied to the management accounts. The quantitative benefits of RA start to fall, as recurring problems have been addressed. The business case for monitoring has subtly shifted from finding holes in revenues to proving there are no holes in revenues. At the same time, the interactions between the RA team and the rest of the business have significantly increased and changed in nature. There is now a ready and amicable two-way interaction with all parts of the business, with RA principally being seen as a business advisor and partner instead of as a policeman or auditor (the RA team also retains the latter role, but the emphasis has changed because fewer issues are being identified and more time is spent on prevention than cure). The goal has become one of designing-in controls and integrity into the business. The rest of the business will voluntarily engage with RA with a view to getting not just their "sign-off" on changes and product launches, but also getting their input on how best to avoid mistakes before they occur. Because the RA team is engaged in advance of any changes to the business, their understanding of the coverage of controls is complete. This, in turn, means they promptly deploy new controls in response to change, maintaining control coverage without any hiatus. A virtuous circle develops between the reactive and proactive wings of the RA team, with the proactive wing planning needs and working hands-on with the business to prevent errors, and the reactive wing providing objective data to substantiate what errors have still occurred and hence effectively giving feedback on the successfulness of proactive work. This in turn is fed into a learning cycle that leads to better anticipation in future. All of this occurs with the direct visibility and support of a C-level executive, who has bought into the mission for RA and appreciates the relative maturity and sophistication of RA in their business compared to the norm.

At maturity level 5—the optimizing level, we reach the pinnacle for RA in the business. Not only the reactive but also the proactive wings of RA have reached an optimal level of delivery. The proactive mode has reached the point where engagement with the RA team drives less advantage because other parts of the business have embedded the principles and goals of RA within their cultures and activities. Reactive monitoring also becomes more decentralized now, as some of the pioneering work of the RA team is handed back to operational units who monitor their own performance and who only need to report back to RA when rare exceptions occur. As a consequence, RA evolves into a complementary and proper element of business-wide risk management, oriented around the collation of data to measure performance and proactive techniques including process mapping and statistical techniques like Six Sigma to improve performance. As such, the boundaries between RA, cost assurance, margin maximization, and even things like service assurance and customer satisfaction monitoring become blurred. The RA department becomes more highly skilled, but fewer in number, with its staff performing the role of a watchdog or guardian, responsible for anticipating and planning for future needs, most particularly with regards to controls. The routine monitoring work is farmed out to other departments or is outsourced. The need for a particular C-level executive champion lessens slightly as all C-level executives recognize the need for RA throughout the business, and take a holistic view of how to deliver it.

Five Aspects

Because there is a lot to the unfolding story of increasing maturity, it is critical to break out the components of maturity into manageable and coherent themes. That way, progress can be more usefully tracked. We can call these themes the "aspects" of RA. There are five aspects:

1. Organization: These are formal elements of how a business is organized, including who reports what to whom, how well defined RA jobs are, how different departments interact on a procedural and decision-making level, and the degree of executive sponsorship.
2. People: This dimension is about the human resources dedicated to RA, and the people around the business who need to support RA. It concerns their skills, education, experience, and attitudes.
3. Influence: There is a big difference between formal reporting lines and the actual ways influence can work in a business. Much of the interaction around RA is informal or hard to systematize, because it concerns errors or issues that may be unique and not repeated. This dimension addresses how the dedicated RA staff work with other functions in the business. It concerns actual relationships and not just what is presented on an organization chart. Within this context, influence concerns who takes a lead with various key activities and the perceived satisfaction with the mission and work of RA.
4. Tools: The extent and sophistication of automation for RA is a very tangible indicator of maturity, but can be given too much significance as a result. Raw power must be considered alongside other considerations like how efficient the tool is and how well designed for its purpose.
5. Process: RA concerns processes and is itself a process when viewed at the highest level. To increase maturity, there must be a striving for process improvement.

Key Concepts

Before we go further, we should outline some of the key thoughts on how the five-level, five-aspect maturity model should be applied in practice.

There is no leapfrogging of steps. A provider at level 2 must next plan to get to level 3. It should not plan to jump ahead to levels 4 and 5. That is because there is an education process that occurs as a business moves up the levels. Different lessons are learned at each stage of maturity, and this cumulative education process is vital for altering the business culture and embedding RA within business operations.

Moving too quickly risks leaving RA as a superficial graft on to the business, which may later be rejected or dispensed with.

The maturity of each aspect can and should be measured separately. Distinct tactics are needed to improve each aspect, so it is appropriate to assess each aspect in isolation. The maturity of the CP is no better than its least mature aspect. For example, an organization with tools that score high on the maturity scale will not get good consistent use from them if the processes around them score low on the maturity scale.

To improve consistently, it is hence important to focus improvement efforts on the weakest aspects, bringing them up to the same level as stronger aspects. Then the plan becomes one of moving up across all aspects at the same pace, one level at a time.

An Illustrative Example

Figure 4.3 shows the results of a hypothetical appraisal of a CP's RA maturity. In this example, the provider has reached level 5 for organization and tools, is at level 3 for influence and process, and at level 2 for people. The threat to RA in this CP is that its success in attaining optimal organization and tools may be undermined by its relative weaknesses, especially with respect to its people. The top priority would be to raise the maturity level on the people dimension to level 3, and then to drive forward the maturity of the people, influence and process dimensions together. Because there are large differences in the maturity of each aspect, it is likely that the

Figure 4.3   Example revenue assurance maturity score.

maturity of the organization aspect or even of the tools aspect may fall over time; this should not be overly resisted because further investment in these areas may give smaller returns than investment oriented toward improving the weaker aspects.

Some Background

The RA maturity model is inspired by what is now called the capability maturity model integration (CMMI) developed by the Software Engineering Institute at Carnegie-Mellon. It follows the 5-level staged representation in the CMMI, which was also used in the original capability maturity model. This step-by-step representation was considered the easiest form of representation for communicating and organizing the ideas in the RA maturity model. There are many analogies between RA and software engineering, not least of which are that they both involve information technology, the finding and reduction of faults, that faults occur due to errors made during change, and that faults may be difficult to either anticipate or detect, but in general we should expect there to be faults. Hence, there are suitable analogies as to how to tackle these challenges and move up the maturity scale. In either sphere, faults are more likely if quality control is not incorporated as an essential element of design, production, and change management. Also, faults are more likely if there is an absence of objective measures of the frequency and severity of faults.

It was necessary to keep the RA maturity model simple in order to be realistic about the typical amount of time that practitioners have available to do strategic planning. Nevertheless, it retains the sophisticated underpinning of the CMMI, and draws upon the experiences learned with that technique. The CMMI itself is supported by more fundamental ideas on how to improve business performance, including the concepts of the plan-do-study-act (PDSA) cycle pioneered by the American statisticians Shewhart and Deming. The RA maturity model was constructed on the basis of the practical experience of real operators and service providers, but was also mindful of its theoretical foundations. These foundations are supported by more empirical data and a longer history of practical experience than is available when solely examining RA in CPs.

Maturity Need Not Mean Ever-Increasing Resources Dedicated to Revenue Assurance

One discussion point in more mature providers is that some are finding they are reaching the "end" of their work and can scale back their RA teams, while others are arguing they need ever more resources to be dedicated to RA. How can both scenarios be true at the same time?

The maturity model is based on the idea that, as you reach an optimizing level of performance, it should be possible to reduce expenditure on RA. In short, the benefits start to diminish, and performance becomes more efficient, so a lower level of investment is justified. The basic growth curve for an RA department in a business is like that of the Product Life Cycle.

- There is an introductory phase when the business first starts doing RA, and during this phase quick wins tend to be realized.
- There is a period of growth that steadily flattens off as the coverage of controls is expanded to all revenue streams (and cost streams, too, if these are included in scope).
- When the maximum scope is realized, all of the quick-win benefits will be realized, plus many of the root causes should have been addressed. There are hence diminishing returns from monitoring. Monitoring should also become more efficient due to increased automation and because staff reach the top of their learning curves.
- The business receives smaller gains from RA, so the resources provided will be reduced down to a more appropriate level for a permanent and stable level of business control.

Why, then, do RA departments often still need to grow? There is a straightforward explanation: the scope that was set for RA is very likely to change over time. Many RA teams start with a narrow remit and are happy to stay narrow because of the limited resources at their disposal. As it becomes successful, an RA team that originally started working on one small field will often expand its horizons. For example, postpaid assurance expands to cover prepaid products. A scope that only covered retail revenues would be expanded to cover wholesale and interconnect, and to include costs as well as revenues. New products are launched, creating the potential for the dedicated RA team's scope to be increased. Sometimes new businesses are acquired. The expansion of scope will drive new investment into RA at a time when older activities are declining in significance. This may be a natural win-win for the business and its RA employees: at a time when its best and most experienced staff are facing diminishing returns as a result of their success, they are given new challenges that will enhance their career, maintain their interest, maximize the value they add, and make retention easier.

The same growth and maturity curves should be evident for the work performed under the extended scope, but obviously, if work starts later, the introductory and growth phases will coincide with the declining phase for the original scope. When applying the maturity model, it is worth considering whether it would be more appropriate to segment the analysis based on the original scope of RA and any subsequent extension of that scope.

Challenges

Some of the challenges in using the RA maturity model revolve around the fact that it is abstract. Abstraction ensures it can provide a single, coherent strategic roadmap, but prevents the model from including detailed and technical recommendations on the specific actions that need to be taken to improve RA.

The benefits of abstraction outweigh the disadvantages. The same model can be used consistently over many years. The model also enables comparison between very different business units. This is particularly useful for groups that use the maturity model to assess a number of operating units in different countries. Finally, the model ensures focus remains on the strategic imperatives, not tactical or operational detail.

It is worth noting that the model is agnostic about who should do RA. People working in a department called "revenue assurance" may do it. It may be done elsewhere, scattered around many departments. Chances are that in most CPs the truth is a bit of both. RA may also be outsourced. It is important to keep in mind that not all RA need be done by people with the words "revenue assurance" in their job titles or by staff who work for a department called "revenue assurance."

Another key point incorporated into the maturity model is that there can be conflicts of interest between the RA department and the interests of the business as a whole. This is no surprise; we live in the real world. A model that says there may be a need for reduced investment in RA as it reaches optimal maturity will not suit the personal interests of everybody working in RA. This makes it important to emphasize that the model is written with the interests of the whole business in mind, not just the interests of an RA department. There are many reasons why this may be hard to accept by individuals working in an RA department. In particular, they may get used to the idea that they act in the company's interests and have to fight for those interests even if the rest of the business disagrees or is ambivalent about them. Nevertheless, one of the goals of maturity is increasing efficiency, and if efficiency is delivered through outsourcing or handing over operational tasks to other departments, this may conflict with any protectionist attitudes in the RA department. This kind of conflict of interest is inevitable given that senior RA staff are expected to simultaneously run a team that executes low-level assurance tasks whilst they also stand back and advise the business on how best to protect its interests. It is recommended that, for the most reliable results, the maturity model be applied in a completely objective fashion, from the viewpoint of the business as a whole.

Obstacles to Improving Maturity

There are some common causes of conflicts of interest, which in turn become obstacles to improving maturity.

Automation should reduce the staff time that needs to be spent on repetitive monitoring tasks. However, staff may lack the skills, knowledge, or motivation to take on new tasks to compensate. They may fear losing their job. The result can be an inefficient distribution of effort, with too much time spent on areas that are well under control, and too little on those areas that most need them.

Cultural change is one of the signs of increasing maturity. As business cultures become more enthusiastic about RA, and more time and resources are engaged in identifying and preventing flaws, the policing and awareness-raising role of an RA department will yield diminishing returns. It will become appropriate for the RA department to take more of a business advisory role, working with other units and sharing credit for the benefits delivered. However, it can be hard to change personal attitudes if the policing mentality is deeply ingrained.

Many RA teams find it beneficial to employ people with software and database technical and development skills, particularly during the earlier stages of maturity. Applying the 80/20 rule, simple tools developed in house may demonstrate the value of RA at much lower cost than more sophisticated off-the-shelf deployments. The skills of in-house developers may continue to be in demand. They may effectively transition into the role of in-house system integrators that assist with deploying data feeds needed by tools purchased off-the-shelf. However, the developers can also become obstacles, especially as development shifts toward support and maintenance. There may be a reluctance to properly productionize their developments, or to make developments sufficiently transparent to open up the choice of support. These people may also be highly resistant and even scornful of the benefits of preventative controls.

Another concern with developers, but also relevant to all staff dedicated to RA, is that a lot of effort is expended on learning: learning about products, learning about technology, and learning how the business works. This knowledge is at a premium. Staff may be more highly motivated to acquire this knowledge than to share it with their colleagues. It is vital that any team adopts a collegiate approach to knowledge sharing, supported through formal and informal learning activities, to prevent the inefficient concentration of knowledge in just a few people.

Reappraising Maturity, Scope, and Purpose

Maturity can be gauged as often as people want; once a year is appropriate for a strategic review based around maturity. It might be a good idea to schedule time at the appropriate point after the financial year-end, or in advance of the annual budget setting, or during the annual Christmas IT freeze. The appraisal is most useful when considering improvements, or lack of improvements, from one year to the next. The appraisal is particularly useful when performed alongside the review of the RA department's scope and purpose. Because the maturity model is based on the best RA interests of the organization as a whole, it serves as a reminder that the RA department does not have sole ownership of all RA tasks. The appraisal can be an objective counter weight to potential drift in the mission of the RA team. Drift can occur in two ways: by accumulating new tasks that are not within the scope of RA or which would have been better realized elsewhere, or by holding on to old tasks that should be farmed out. Performing the maturity appraisal serves as a reminder about these options and also sharpens communication to executives about how the work being performed by the RA team corresponds to the RA needs of the whole business.

The Maturity Model Itself

The backbone of the maturity model is the questionnaire. It greatly deepens the analysis of what makes for maturity in RA. However, it is quick and easy to complete. The questions can be answered by someone with good familiarity with the CP being reviewed, and general awareness of RA. There is no need to collate special data or metrics or to have expert or detailed knowledge of the workings of the business or of RA.

There are 25 questions in the questionnaire in total. Each aspect of maturity is covered, there being five questions for each aspect. For example, the questions numbered 1–5 relate to organization, while 5–10 relate to the people aspect. The answers for each aspect should be separately collated and averaged. The idea is to consider each aspect on its own, to determine where to focus effort to increase maturity.

Each question comes with five example answers, where each answer corresponds to one of the five levels of maturity. The idea is to select the answer that most accurately describes the circumstances of the CP being appraised. There may not be a perfect match to any of the answers; the idea is to select the closest match of those available. In many ways it is better to go with the first answer that feels right, rather than spending a long time contemplating which the best match is. If it is very hard to choose between two answers, it is permissible to pick a halfway score. For example, if the reality appears to be equally close to the answers for both level 3 and level 4, the raw answer to that question could be recorded as 3.5.

The complete questionnaire is presented at the end of this article. You may want to flick ahead and take a sneak peek, but come back and read the remainder of the article before you start using it in earnest.

Worked Examples

The way to complete the questionnaire is best explained by doing some worked examples.

This is the fourth question in the questionnaire, and relates to the organization aspect of maturity: "Are revenue assurance duties segregated to avoid conflicts of interest or potential for bias?"

In brief, the possible answers are

1. No segregation.
2. Some segregation due to tasks being spread across functions.
3. Segregation between the RA team and other teams, but not within the RA team.

4. Segregation both between the RA team and other teams, and within the RA team.
5. The RA team works with other areas to get the right balance of synergy and segregation.

Five detailed answers are provided in the questionnaire, each one corresponding to a maturity level from 1 to 5. For brevity, the answers given above are only summaries. If you take a look at the questionnaire, you will see the answers go into more detail.

In this example, let us suppose that answer 3 is the best match to the actual circumstances of the CP.

Now suppose we have completed the answers for the other four questions relating to the organization aspect. We take the numerical value of the five answers and calculate the mean average. So if the five answers were 3, 4, 3, 2, and 5, we add the answers together to get 17, and then divide by 5 to get 3.4 as the mean average. 3.4 is the maturity score for the organization aspect of the CP being appraised.

We complete the questionnaire and calculate the scores for each other aspect: people, influence, tools, and process. We get five separate scores because the idea is to focus more attention on those aspects that yielded the lowest scores, with a view of bringing them up to the same level as the top aspects.

You may also want to calculate an overall average by taking the five scores for the five aspects and working out the average across all of them. This has little value as a decision-making tool, but may be useful for succinctly communicating progress from one year to the next.

Applying Results in Practice

To improve performance in one of the more poorly performing aspects, the obvious thing to do is to focus on the questions that returned the lowest answers. The strategic goal should be to improve performance in relation to those questions that yield low answers.

To give another example, consider a CP that scores 4.3 for organization, 2.3 for people, 4.1 for influence, 3.8 for tools, and 2.2 for process. This provider should seek to improve the maturity relating to the people and process aspects. Only when they have been raised to a similar level to the other aspects should the priority shift to putting more emphasis on improving other aspects with a view to attaining an optimizing level of maturity.

To reiterate, it is important to remember that the stepped model is not meant to be linear. The actions to get from level 1 to 2 are different for those to get from 2 to 3, 3 to 4, and 4 to 5. Completing each level is a prerequisite for getting reliable progress later on. This distinction between the levels, with no bypassing or leapfrogging of levels is a vital difference between this maturity model and less sophisticated (but often overcomplicated) strategic models for RA.

Maturity and Proactivity

What is the relationship between maturity and proactivity? The term proactive is used a lot in RA circles, but what does it mean? Unfortunately, it means different things to different people. To avoid confusion, use the proactive, active, reactive (PAR) nomenclature, as follows:

- Proactive means taking action before the event that could cause a leakage occurs. Designing a user interface to stop the entry of invalid reference data would be an example of proactive RA.
- Active means taking action after an event that could cause leakage, but before the leakage is realized with a real impact on the business. For example, a daily review of rating system suspense files might catch a problem with reference data, allowing it to be fixed and the files to be reprocessed in good time so that all calls are still billed correctly at the end of the month.
- Reactive means to take action after a leakage has taken place and impacted the business. For example, a trend analysis is performed after the billing run, and this indicates that revenues were lower than expected when compared to the previous month. The reason for the variance is investigated and found to be an unresolved fault with reference data. This caused a high volume of events to fall into suspense, where they remained unrecycled when the bill run was executed.

The PAR definition helps to cut through marketing hyperbole and enable staff to precisely focus on how best to deliver better RA. For example, being proactive is not just a loose way of saying you do things more quickly. Contrary to what is reported (or deliberately overlooked) in advice from some quarters, RA controls need not only be detective in nature, waiting for something to go wrong and then deciding how to deal with the problem. RA can also implement preventative controls, stopping problems before they occur.

There are two ways that RA can improve its performance and reduce leakage by tackling its turnaround time. It can seek to reduce reaction times, moving controls from reactive to active. It can also supplement or even replace reactive and active controls with proactive measures to prevent errors that could lead to leakage.

These ideas about moving from an essentially reactive approach, to one that is active, and then broadening to become proactive, underpin many of the descriptions used for maturity in the maturity model. The mindsets for making reactive controls work more quickly and for implementing proactive controls are quite distinct, but they can be pursued in a way where they complement each other. This again harks back to Deming and the PDCA cycles that are the intellectual forebears of the Capability Maturity Model and hence of the RA maturity model. Deming sought to reduce reliance on postproduction inspection by building quality into production processes. The same thinking can be equally well applied to

Table 4.1   Maturity and Proactivity

Reactive		Active	Proactive	
Initial	Repeatable	Defined	Managed	Optimizing
Ad Hoc Projects	Recurring Monitoring	Real-time Alarms	Design Reviews	Risk Management

data as it is to manufactured goods. This maturity model does not seek to do away with inspection on a mass basis, but it does seek to reduce reliance on inspection by complementing it with a drive to build-in quality—that is, to build-in revenue integrity—into the systems and processes of the CP.

Table 4.1 gives a simple representation of how the stages in evolving maturity link to an increasing shift toward active and proactive RA. The table reads left to right, from a low level of maturity and high degree of reactivity to a high level of maturity and high degree of proactivity. The bottom row captures what major new improvement since the previous maturity level is necessary to support the improved timeliness of RA activities.

At the first and second levels of maturity, all work is reactive in nature. Ad hoc projects to find and fix suspected issues and the subsequent implementation of regularly recurring monitoring activities are based on the idea that errors will occur and these need to be detected after the fact. In the defined level of maturity, the business will implement controls that are active in nature—at least for post-paid services—in that controls will alarm in real time and hence enable fixes to be implemented quickly enough to avoid leakage from taking place. Also, this is the stage where the organization starts to take the lessons learned and reapply them to reduce the risk in future. This tendency significantly increases in the latter stages of maturity, the managed and optimizing levels. In these stages, design reviews and risk management are incrementally added to the portfolio of RA activities. They complement the active and reactive controls in place, making errors less likely. These controls, in turn, provide data on whether preventative steps are being successful at anticipating and reducing the likelihood of error.

Summary

So to summarize the trends we expect to see in increasing maturity:

■ To get to level 2, the main challenge is to make RA work regular and repeatable. A sign of success is that an RA activity should not be adversely impacted when a member of staff leaves.

■ To get to level 3, the coverage in detective controls nears 100% and there is the building of a repository of knowledge and a series of real-time alarms, typically concentrated in a centralized RA team.

■ To get to level 4, leakage is not just monitored but is also consistently quantified in financial terms, allowing the benefits of RA to be readily reconciled to the accounts. Monitoring becomes more efficient and handovers for resolving problems are all well defined. Anticipation of the causes of revenue leakage becomes a significant new wing of the strategy to mitigate the risk of leakage.

■ To get to level 5, the optimizing and highest level of maturity, risk management has become so embedded that the proactive and reactive elements of RA now work as a harmonized PDSA cycle. Efficiency of assurance is important because returns have significantly diminished. RA also becomes increasingly decentralized, as it becomes embedded in the thinking and culture throughout the organization.

The New Questionnaire

Please refer to p. 57 for the Maturity Questionnaire (Table 4.2).

Revenue Assurance Is a Silo!

Ashwin Menon

The way revenue assurance (RA) is currently being viewed by most communication providers tends to give me the impression that RA is well on its way to being siloized. Having silo systems with no interaction/feedback moving through the value chain is one of the reasons for revenue leakages in the first place. Like, for example, the switching team introducing a patch that causes a difference in the normal product stamping in the downstream systems. However, the news of this patch being applied is suppressed (perhaps because it might be a quick fix for an earlier issue that was not highlighted). As a result, as the mediation teams have not got any updates, and the incorrectly stamped xDRs are summarily rejected.

Imagine a scenario where an RA department does not disseminate findings immediately, but instead waits for a weekly review. Imagine, the RA department mails another department their findings, but due to ineffective workflow management, the issue is suppressed. Suddenly the switch and mediation teams are face to face with issues that could have effectively been culled at a grass-root level, but because of linear propagation, the patch has been transferred to other elements, which has effectively bloated up a minor issue.

Though this is not in the truest sense a siloizing of the RA department, it becomes critical for operators to not only have a good RA methodology or a good team, but it also becomes important to implement a solid, leak-proof workflow and processes that govern the entire life cycle of a discrepancy, by which I mean,

Detection → Investigation → Root Cause Identification →
Qualification → Quantification → Resolution

Table 4.2 Maturity Questionnaire

No.	Aspect	Question	Level 1: Initial	Level 2: Repeatable	Level 3: Defined	Level 4: Managed	Level 5: Optimizing
1	Organization	Who in the business effectively takes the lead for driving the improvement of RA?	A lone individual who has chosen to make RA part of his or her job	Several disconnected individuals, each taking responsibility for RA in areas that relate to their job	The manager who heads the team dedicated to performing RA	A C-level executive	Responsibility is distributed amongst C-level executives to the point that it is difficult to say who takes the lead with improvement
2	Organization	Is there a permanent team or teams dedicated to RA?	No; what RA there is, is done on an irregular basis by a team with a different nominal responsibility, such as Internal Audit	No; some RA tasks are performed on a regular basis alongside other responsibilities by people working in a variety of areas like Billing or IT Operations	Yes; permanent staff are dedicated to RA, but they ultimately report to a manager where there may be a conflict of interest with their other responsibilities, such as Billing	Yes; a dedicated RA team reports to a C-level executive and performs the bulk of RA activities	Yes; the team reports to a C-level executive but mostly oversees RA activities distributed around the business, meaning that some dedicated RA roles are employed in other functions

continued

Table 4.2 Maturity Questionnaire (continued)

No.	Aspect	Question	Level 1: Initial	Level 2: Repeatable	Level 3: Defined	Level 4: Managed	Level 5: Optimizing
3	Organization	Is responsibility for RA held by a central function or distributed around the business?	Centralized because only the lone champion takes responsibility	Distributed between the multiple functions that each perform some RA tasks	Heavily centralized in the dedicated RA team	The centralized RA team is complemented by an increasing number of distributed stakeholders within the business	A small centralized team retains responsibility for oversight but most low-level recurring tasks are distributed around the business
4	Organization	Are RA duties segregated to avoid conflicts of interest or potential for bias?	No segregation as the lone champion does RA alongside his or her other responsibilities	Staff in the several departments that perform RA tasks do them alongside other regular tasks, but distribution of responsibility means some de facto segregation	The dedicated RA team imposes functional segregation by taking on the bulk of tasks, but segregation within the team may be weak	Segregation occurs both at the level of having a separate, dedicated RA department and there is segregation within the RA department	The team that oversees RA is responsible for working with other functions to ensure the right balance between operational synergy and segregation of duties

5	Organization	How flexible is the organizational design when it comes to adapting RA for changing business models and other factors that alter the priorities?	No flexibility in structure and the lone champion is as flexible as his or her normal job duties, skills, knowledge, and personality permits him or her to be	Flexibility exists in the form of ad hoc and temporary projects and initiatives, though it can be slow to get approval for these	Appointment of permanent RA staff with specific duties reduces flexibility unless efforts are made to appoint and reward multifunctional staff with an open remit	A good flow of information between the RA department and other functions encourages a flexible approach that responds to changes in priority	The central RA team focuses on looking ahead and ensuring the business develops the capabilities needed to maintain assurance in the midst of change
6	People	What skills and experience do RA staff have?	The skills and experience reflect the lone champion's background	There are a wide variety of skills between the disparate individuals performing RA tasks, but little experience in RA itself and many gaps in skills needed	The skills and experience of the RA team are generally biased to the function that RA sits in; for example, if RA is a subset of Billing, then staff tend to be experienced in billing but not other areas	There is a good blend of skills, both technical skills and soft skills, exhibited by the central RA team	The core RA team is highly flexible but instead of seeking to master all the areas of skill required, they work interactively with specialists around the business

continued

Table 4.2 Maturity Questionnaire (continued)

No.	Aspect	Question	Level 1: Initial	Level 2: Repeatable	Level 3: Defined	Level 4: Managed	Level 5: Optimizing
7	People	How are RA staff trained?	All training comes in the form of encouragement from the lone champion and learning from the experience of trying to do new things	The involvement of several function areas opens up opportunities for sharing of knowledge and cross-fertilization of ideas on how to do RA	A demand for certification and prepackaged classroom training reflects naivety and a limited perspective on how to develop staff successfully for a varied, hybrid role	There is a strong emphasis on in-house exchange of knowledge within the RA team and between RA and other functions; this occurs on both formal and informal level	Sharing knowledge within the business is complemented by a sophisticated commitment to sharing with and learning from other CPs
8	People	How are RA staff recruited and retained?	The people who do RA tasks do so from a mix of personal commitment and the ability to balance the additional workload with other responsibilities	Staff are typically assigned to RA tasks because of similarities to their previous and current duties	People are recruited from both within the business and from outside; a standard checklist of requirements for new staff may mean a failure to broaden the skills within the team	Recruitment is less important as more work is automated; the focus shifts to retention by upskilling and giving staff varied challenges; the more routine jobs are offered to graduates and school leavers	The highly skilled members of the smaller core team are developed to be future executives and recruitment focuses on individuals with the relevant potential and ambition to be a business leader

9	People	How does the business cope with short-term demands for extra human resource to cope with specific one-off problems?	It is unlikely that additional temporary manpower is engaged unless the lone RA champion gets the support of other functions for a mutually beneficial project	Staff from across the business may be reassigned for short periods to cope with specific issues	The employment of permanent dedicated staff may discourage approval of temporary manpower, but the RA team becomes better at making a business case for temporary contractors	There is a mixed and balanced strategy of using outsourced, consultant, temporary contract, and redeployed in-house manpower to deal with one-off issues	Better planning and more proactive management of RA reduces the need for temporary increases in manpower
10	People	For staff who have an RA responsibility but are not dedicated to RA, how is their contribution to RA monitored and appraised?	The contribution made by staff performing RA tasks is largely overlooked	Line managers in a variety of functions take different and inconsistent approaches to informally monitoring the RA activities of people in their area	Focus on the dedicated RA team reduces attention for staff doing RA tasks elsewhere in the business; however, the RA team does start to encourage recognition for the RA efforts of other staff	Recurring RA activities are formally written into the objectives for staff in other functions	In addition to line managers appraising formal RA objectives, the RA team engages in succession planning by looking to second staff who show potential

continued

Table 4.2   Maturity Questionnaire (continued)

No.	Aspect	Question	Level 1: Initial	Level 2: Repeatable	Level 3: Defined	Level 4: Managed	Level 5: Optimizing
11	Influence	What influence does the goal of RA have on the decisions the business makes?	The lone champion exerts influence as and when possible, which is rarely	The threat of embarrassment motivates functions to deal with the most serious leakages, but there is little incentive to look for previously unidentified issues	The RA team pushes for changes in decision making with variable results depending on how well they are aligned to the overall business priorities	The RA team and their sponsoring CxO exert influence on a wide range of business decisions but do so in a way that is well aligned to overall priorities, thus minimizing conflict	RA goals are sufficiently ingrained in the business culture that they become inherent to all decisions

12	Inf uence	What is the awareness of RA across the business?	RA is unknown to anyone but the lone champion and a small group of colleagues	Staff who undertake specific and relevant tasks are aware of their importance, but may not think of the tasks as being "revenue assurance" or part of a wider series of objectives	The RA team increasingly raises awareness of RA when dealing with other functions in the business	The RA team engages staff on a broad basis, ensuring that there is good awareness of the need for RA	Awareness of RA is made an element of the on-boarding process for new staff and is reinforced both formally and informally at regular intervals
13	Influence	How do staff performing RA activities get the information they need?	If staff do not already have the information as part of their existing role, they are unlikely to get any further information for the purposes of doing RA	There is a limited exchange of specific information between staff in disparate functions that have common RA goals	The RA team specifically requests information, pulling it from systems and people; little information is pushed to the RA team	Other functions are likely to provide useful information to the RA team without waiting to be prompted; the RA team has good visibility of most relevant data	The RA team is in continuous communication with other functions, sharing information and dealing with issues on a "by exception" basis

continued

Table 4.2 Maturity Questionnaire (continued)

No.	Aspect	Question	Level 1: Initial	Level 2: Repeatable	Level 3: Defined	Level 4: Managed	Level 5: Optimizing
14	Influence	How consistent is the measurement of RA benefits with the way performance is measured for other business functions?	There is no measurement of RA benefits	RA benefits are not clearly broken out and are only sporadically reported within the context of the performance reporting of the functions that engage in RA activities	To get fair treatment, the RA team demands it is assessed differently to other functions; the benefits it reports may not be reconcilable to other reporting, such as the management accounts	The RA team reports on benefits delivered by it and by other functions engaged in RA activities; these are reconciled to the management accounts	Consistency in measuring performance versus the business' objectives is of paramount importance to the RA team
15	Influence	Is the justification of RA activities economic in nature?	Yes; the lone champion seeks to provide evidence of the economic benefits of RA activities	Yes; functional heads invest manpower and other resources into RA on the basis of how much it supports the budgetary targets of their function	Yes; the RA team primarily argues that it delivers economic benefits, even though this creates tensions when measuring the value of some of the proactive work it starts to do	No; the benefit to the bottom line is just one of the several quantitative and qualitative justifications for RA	No; there is perfect synergy between RA and the overall business objectives, meaning it no longer has a clearly separable justification

16	Tools	What is the type and blend of tools available to perform RA?	Ad hoc adaptation of standard PC software	Limited reuse of other software and equipment not originally purchased for RA purposes	Tools are primarily focused on automated detection of leakages and reporting on the leakages detected; limited or no use of workflow tools	Tools are used to detect leakages and are integrated into business processes to enable straightforward workflow for recovery of leakages and management of incidents	RA tools become less separable because the relevant capability is often integrated with other automation such as service assurance or business intelligence
17	Tools	How much manual intervention is needed to support the tools?	Only basic elements of processing are automated and a high degree of user intervention is needed at all times	Functionality is narrow and rigid, meaning there are a lot of manual steps to adapt, prepare, import, and analyze data	Basic data-intensive recurring tasks can be performed with relatively little user involvement; staff focus more on adapting technology for more unusual issues	Users spend time configuring and optimizing systems but manual intervention is not needed for basic processing tasks	Sophisticated corporate-wide tools for business intelligence and workflow provide RA with a ready-made foundation that their dedicated tools integrate with

continued

Table 4.2 Maturity Questionnaire (continued)

No.	Aspect	Question	Level 1: Initial	Level 2: Repeatable	Level 3: Defined	Level 4: Managed	Level 5: Optimizing
18	Tools	Are RA requirements considered when purchasing systems that are not specifically for RA?	No	Narrow and specific RA requirements are factored into the purchase of systems with the most obvious relationship to RA, such as mediation or billing	RA requirements are tabled for procurement of any system as long as the RA team has identified the potential benefits at an early enough stage	The RA team is sufficiently forewarned about technology transformation that they can routinely build in their requirements wherever relevant	RA requirements are routinely anticipated by other functions who then confer with the RA team to ensure their needs are satisfied
19	Tools	What is the level of investment in tools relative to the investment in people?	No investment in tools; staff spend time adapting what they have	Some staff are engaged in developing tools; in-house development and adaptation is preferred to purchasing off-the-shelf solutions	Investment in tools increases to expand the coverage of detective controls, without there being any impact on the human resources deployed for RA	Investment in soft tools designed to improve productivity and manage workflow leads to a reduction in manpower needed for routine tasks	The central RA team utilizes a highly sophisticated alarming and workflow architecture to oversee the work of other functions

| 20 | Tools | Can the tools used by RA be used for other purposes? | The tools used for RA were not intended for the purpose of RA and hence are already better suited to other purposes | The functionality available for RA tasks is narrow and does not lend itself to other activities | RA tools have been developed and procured with a tight focus on RA requirements, with little consideration of the potential to use them for other goals | The RA team use imagination to reuse tools and data to support activities that fall outside of the traditional RA scope | The tools used for RA are inherently multifunctional |
| 21 | Process | Is there an RA framework? | Work is ad hoc and chaotic, following no systematic process | Processes are basic and repeatable, but they are narrow and relate to particular objectives with no attempt to address wider goals | The RA team formulates a basic approach to detecting revenue leakage that is incrementally rolled out until it has been used to address the team's full scope | The RA team engages other teams in a collective framework to both detect and prevent leakage | The RA framework is an integrated component of the business' risk management framework |

continued

Table 4.2   Maturity Questionnaire (continued)

No.	Aspect	Question	Level 1: Initial	Level 2: Repeatable	Level 3: Defined	Level 4: Managed	Level 5: Optimizing
22	Process	What is the basis for designing RA processes?	Processes are entirely ad hoc and hence reflect the instincts of the person executing the process	Processes are low level and procedural in nature; they have become routine but little consideration has been put into their efficiency	The RA team defines its processes and tries to do so for other functions; there is a tendency to start from scratch with little reference to established techniques for process development	The skills of the RA team incorporate process improvement, meaning that RA processes are designed to be efficient and to reflect industry best practice	All business processes are designed to be efficient, and because of this RA processes are highly integrated and synergized with other activities
23	Process	To what extent are RA processes subject to continual improvement?	There is no attempt to improve RA processes, as they may not be executed consistently enough to make it worthwhile investing effort in improvement	Processes tend to remain unchanged unless there is an obvious fault with them	Processes are defined in quite a theoretical manner, meaning there is a lack of feedback and a lack of internal review that would identify opportunities for process improvement	Processes are regularly reviewed with a view to ensuring they are effective and efficient	The RA team's focus shifts toward optimal management of the relevant processes extant throughout the business

| 24 | Process | How are RA processes measured and monitored? | RA processes are not measured or monitored | Line managers oversee staff performing RA tasks and look at relatively simple metrics relating to process performance to ensure that the processes have been executed | The RA team dedicates significant resources to measuring and monitoring but the focus is on highlighting leakages and calculating benefits, rather than improving the processes | The RA team engages with other functions to ensure that relevant processes are measured and monitored as part of routine operations and with an intention to improve performance | The purpose of the RA team is largely to monitor and measure processes rather than to scrutinize detailed data |
| 25 | Process | What is the process for evaluating and setting priorities? | Not applicable | Priorities are set by informal dialogue between departments, usually without referring to data to justify the decisions made | The RA team makes use of data to set priorities based on a quantitative evaluation of the impact of leakage on the bottom line but may lack visibility of overall business priorities | The RA team works with other functions to agree priorities based on quantitative data and qualitative but objective judgment | Priorities are set on the basis of extensive quantitative and qualitative data with a keen sense of alignment to the overall business priorities |

continued

As someone in charge, I should be able to track, at least at a high level, the status of a particular issue with respect to the above flow.

I heard a little story about Japanese fishermen throwing small sharks into fish tanks so that the fish are kept in a constant state of activity, and when they reach shore, the fish would be surprisingly fresh. The point I am trying to make is that for true RA coverage, it is not enough to simply identify issues but it is also critical to be able to track it to closure. If the fish represent the discrepancies that were discovered, the shark would allude to vigilance on the part of communication providers in being able to track progress/resolution. Else, all we get are rotten fish—also known as lost findings.

Using Variances to Set the Scope for Assurance

Eric Priezkalns

Does revenue assurance (RA) generate revenues? Does it generate cash? Does it generate profits? I do not think so. Or, at least, it generates these benefits no more than sales people generate them, or the people who roll out new network infrastructure generate them. Every employee has a role to play in sustaining the business of a communication provider (CP). What confuses the issue is that RA is often the difference between a benefit realized and a benefit "lost." RA comes at the very end of a sequence of events, and the benefits it adds is measured in the terms of the failures of the sequence that comes before it. RA closes gaps. Gaps occur between what should have happened and what actually did happen, yet different RA practitioners in different companies can adopt a very different scope. The priorities for their businesses may vary greatly. Generic definitions of the scope of RA often fail to address the real differences that can occur in scope. Goals stated in terms of "maximizing revenues" or similar phrases do not address what gaps are of interest to the RA practitioner, and which fall outside the practitioner's remit. A better way to understand the scope, and tailor it for each business, is to work methodically through which kinds of gaps, or variances, the RA practitioner is meant to identify and close. Let me explain by beginning with the "core" variance that the vast majority of RA teams include within their scope.

Core Leakage: True Price of Actual Sales versus Total Value Charged

This variance is between the aggregate monetary charges presented to customers and the figures that should have been presented if all goods and services had been charged for at the correct price. This variance is so fundamental that many RA departments treat this not just as the core but also the outer limit of their scope. As such, other RA functions may actually call this the "core" leakage to distinguish it from other kinds of problems they also include in their scope. New RA functions may only aim to address a subset of this core variance to keep their task manageable.

One way to break down this variance would be to analyze it into the three "C"s of capture, conveyance, and calculation. The three "C"s will be discussed in more detail in subsequent articles, but the essential idea is that all processing is either capturing and recording the sale, conveying the data from one place to another, or calculating the monetary value to be charged. There are hence three subvariances within core assurance:

1. The capture variance: the difference between what was actually supplied and what was recorded as supplied
2. The conveyance variance: the difference between the sales records initially captured and those which are eventually presented to the customer for payment
3. The calculation variance: the difference between the correct price for the sales presented for payment and the actual price presented for payment

Other ways to divide the core leakage are based on specific process or transaction flows. For example, there may be a separate analysis of transactions charged on a usage and on a nonusage basis. These are more commonly described as controls over switch-to-bill and order-to-bill, respectively. A common alternative is to analyze the core variance between postpaid and prepaid revenue streams, not least because the end point of each stream is distinct and because the types of errors and controls to detect or prevent them are quite different. Nevertheless, even though the transaction processing varies, these are all examples of the same fundamental variance between the chargeable value of the services supplied and what the customer is asked to pay.

Three Other Fundamental Variances

There are three other fundamental variances that lead up to, or follow on, from the core variance. They are hence often included within (or confused with, depending on your point of view) the scope of RA. Because they connect to the core variance, it is possible to lose sight of where one variance stops and the next starts; distinguishing them is a useful way to ensure clarity because these variances pose quite distinct challenges. The three variances relate to opportunity loss, accounting integrity, and cash collection.

Opportunity Loss: What Could Have Been Sold versus What Was Sold

A simple example of opportunity loss might be that a cell site reaches the point where it supports the maximum number of calls it can. If any other customers within radius of the cell site try to instigate a call, they are unable to do so. As a consequence, the opportunity to make another sale is lost. Because of the kind of data that RA typically interrogates about the volumes of sales captured, it can easily extend its perspective to instances like these, where there is a variance between what could have been sold and what was actually sold. However, care is needed when extending RA's remit to cover opportunity loss. The idea of what could have been sold is very open ended. There are very many kinds of opportunity loss and the RA team is unlikely to be able to analyze them all. Evaluating the financial impact of opportunity loss also requires a degree of speculation. For example, even in the relatively straightforward example of a cell site reaching capacity, we cannot easily judge how many callers would simply wait and place a call at a later time, meaning the sale is delayed rather than lost.

The open-ended nature of managing opportunity loss is best illustrated by the differing attitudes to managing briefcase time, the time it takes to convert an order into a provisioned service that generates revenue. On one hand, this is an aspect of timeliness of processing, and naturally falls within the remit of any RA team that embraces timeliness as a goal. On the other hand, RA teams may think their job begins only when a chargeable service is supplied, so exclude briefcase time as this is to do with the efficiency of serving a customer *before* it is time to raise a charge. However, many RA functions have suitable controls-based and data analytic skills to monitor and improve briefcase time. This can lead to two kinds of benefit. Most obviously, if services are provided sooner, charges are raised sooner and the business benefits from positive cash inflows sooner, and in general will benefit from higher revenues as a delayed start date for a service is unlikely to be balanced by a mirroring delay when the customer decides they want a service disconnected. Second, some customers may churn, while waiting for a service to be provisioned. Indeed, they may churn because of the long wait for a service to be provisioned. Reducing briefcase time means less preactivation churn and hence improved revenues, while lower wasted costs on part-provisioning of customers.

Usually RA departments stay very close to concrete examples of opportunity loss that can be backed by data that the department already has access to perform its capture and conveyance checks. However, once opportunity loss is included in the RA team's remit, there always comes a relatively arbitrary cut-off point for which scenarios they cover and which they do not. To give an illustrative example, a rude customer services representative may be responsible for losing a sale that might otherwise have been made, but it is not the job of RA to get involved in monitoring the quality of training for customer services representatives or their interactions with customers.

Accounting Integrity: Charges Presented versus Charges Accounted for

Presenting charges to a customer in the shape of a bill or an update to the customer's ledger is not the only endpoint for data about a sale. Accounting records should also be updated accordingly. These accounting records may be for external use by shareholders (financial accounts) or for internal use by decision makers within the business (management accounts). The production of accounts is otherwise known as reporting. The task of ensuring that numbers in the accounts match the charges actually presented may be described by the phrase "reporting verification." The role of RA functions in supporting the job of reporting verification may vary greatly. Some take no interest in the task, and it is left to a mainstream accounting function to ensure they are consistent. The task may be relatively simple, or extremely complicated, depending on the degree of automation involved in updating the accounts and the simplicity of the mappings between the data output from systems used for billing and prepaid balance management and the chart of accounts. In other CPs, the RA team will have a hands-on responsibility for demonstrating that the accounts are consistent with the charges presented.

Even when not normally tasked to do so, RA may be included in the analysis of the accounts if unexpected fluctuations against forecast or typical sales patterns lead management to question the integrity of the information it receives. For example, if management accounts show a big fall in revenues for a particular day, this may be indicative of a core leakage issue, and the RA team may be called upon to analyze what happened.

The obligations imposed by the Sarbanes–Oxley Act and other accounting requirements have sometimes confused the role of RA. It should be noted that the RA team's work cannot be said to support accounting integrity, and hence support compliance with requirements like these, unless it has first been demonstrated that there is assurance from the limits of the data that RA does assure to what is presented in the accounts.

Cash Collection: Charges Presented versus Cash Received from Customers

After capture, conveyance, and calculation, collection is the fourth "C" because it is so frequently included in the remit of RA functions. That said, the tasks involved in maximizing the collection of accounts receivable are far from unique to CPs. Conversely, CPs with a predominantly retail and prepaid customer base need controls to manage the collection of payments but if fraud risks are appropriately mitigated they will not suffer the kinds of bad debt "leakage" that are typical of postpaid services. To some extent, a degree of leakage is inherent to any business model where customers are debtors. The data analytic techniques of RA are most helpful when focused on the specifics of understanding the large volumes of transaction data. For example, RA is well positioned to deal with questions like who is very rapidly building up a lot of debt and should have services suspended, or which customers and customer service agents are receiving/giving the most credits.

Other Variances

Building around this foundation, we can see how the remit of the RA function can be extended in different directions, depending on what other variances they explore and manage. The correct priority, of course, is to manage the variances where this will deliver greatest benefit to the business; this will be different in each CP. A brief run-down of peripheral variances that are sometimes included in the scope of real RA functions is provided in the following sections.

Directly Variable Costs Incurred versus Paid

As CPs earn revenues from each other, the flip side is that they incur costs for the services they receive from other CPs. These costs are directly variable with revenues. As such the same control techniques apply when assuring these costs or the associated revenues. Other directly variable costs like revenue share with content providers and with premium rate service subscribers can and should be managed on a similar basis. A Three "C"s type analysis can be applied to such directly variable costs, the only difference being a reversal of the direction of risks, with the aim being to avoid overpayment, while ensuring that potential liabilities are not under measured.

Sales Made by Dealers versus Dealer Commissions

This might be considered a specific example of managing costs with partner organizations. Because dealers have financial incentives to make sales and connect customers, an RA-style of analysis will reduce the risk of inaccurate payments of commissions and of fraud by the dealerships. An RA-style approach to managing dealer transactions is suggested because the most basic form of control would be a matching of the revenues received from new customers signed up by a dealer with the commissions paid to the dealer. Even if all payments are valid and accurate, a comparison of revenues and costs may prompt changes in the structure of commission payments with a view to maximizing the net benefit and ensuring the dealer is correctly motivated to acquire the most profitable customers.

Other Costs Incurred versus Paid

Extending RA's remit to manage costs not directly variable with sales is analogous to the way some RA teams oversee the fourth C of cash collection as well as core leakage. All businesses collect cash from customers, and all businesses make payments to suppliers. The controls needed to manage payments in general are not specific to CPs, but that does not preclude a specialized team like RA being given responsibility for implementing those controls. However, it is rarer to task RA to do this for the very reason that the risk can be mitigated without needing to place the burden on a highly specialized and highly skilled team with detailed knowledge of how specifically CPs work.

Network Infrastructure Available versus Utilized and Billed

This extension of the RA scope is most common in B2B fixed-line providers and cable operators. If network inventory records are not accurate, then an unwary provider may spend money on deploying new assets when old assets were available and unused. In the worst cases, a business customer may benefit from a multitude of lines, some of which are billed, others not billed, and with neither the customer nor the supplier having a clear idea of what is being paid for and what is being supplied. A standard reconciliation-based approach will enable RA to identify underbilling and underutilization of infrastructure. By doing so, they will also improve the quality and efficiency of future investment in the network.

Network Infrastructure Utilization versus Target/Opportunity

This could be seen as a special case of opportunity loss. For a provider like a cable operator or mobile operator, some infrastructure may generate far greater returns than other infrastructure. A cable operator might find they are supplying all the customers on one street, but none a few blocks over. Some cell sites will average much higher levels of utilization than others. Analysis of the data as to which geographical locations generate the best returns can lead to two kinds of financial benefit for the CP. First, the lessons learned may greatly improve decisions about geographical location of future infrastructure investment. Second, the CP may devise marketing promotions and schemes to better sweat current assets. For example, in a neighborhood where there is very low penetration into the customer base, a one-off reduction in fees to connect would generate increased ongoing returns from assets that otherwise are underexploited.

Conclusion

The history of RA can be understood as a gradual extension of the same basic idea: RA works to close the gaps between how transactions are managed and processed in practice and how they should be managed and processed. This construct can be applied in many different directions, though most CPs agree that RA includes managing a common core variance between what services were supplied to customers and what customers were charged for. Beyond this, the relative priorities vary greatly, and so may the way work is distributed around the CP. Some generic control tasks may be handled by RA in one CP, by another function in another CP. This choice in how to allocate responsibility is most obvious when talking about protecting the integrity of accounts or handling the collection of cash. For some CPs, getting the best value from network infrastructure investment may be vital; others may focus on dealer commissions. Each CP should identify the key priorities and frame these in terms of key variances, to avoid confusion about who does what (and what is not being done by anybody). This helps to clarify the scope of the RA function, so it can add the greatest benefits in line with the overall business goals, while avoiding duplication of effort with other departments in the same CP.

Import and Export of Responsibilities as a Tactic to Support the Revenue Assurance Strategy

Eric Priezkalns

One way of strategically understanding revenue assurance (RA) departments is in terms of import and export. This positions the RA department as the fulcrum in a process that delivers RA to the business, but where the role played by the department varies over time.

The RA department may identify an area of weakness, and, resources permitting, start an initiative to address this area. It "imports" responsibility into the RA department, even if other parts of the business already state that they are accountable for the area being reviewed. The job of the department is hence to invent, prototype, and deliver a working solution that addresses the weakness. This solution may be a new kind of monitoring activity. It may also be an improved procedure that will reduce errors and enhance returns; examples would be a better approach to managing suspense or the rationalization of the reason codes used to track customer credits. Once the initiative has delivered a stable solution that is working well on a routine basis, it will usually be appropriate to hand over the control to the relevant operational team. In other words, it "exports" the solution, but keeps a watching brief to ensure it is executed correctly in future. Exporting frees up RA resource for higher-skill work appropriate for a team capable of devising and implementing new controls. In other words, it releases resource ready for another imported initiative.

There is a natural temptation for RA to import at a rate it can cope with, but never to export again. If this happens, the department will get steadily bigger but also will steadily lose strategic direction and a consistent purpose. Over time, it will be unclear why some operating activities are performed by the RA department when other, similar, operating activities are performed elsewhere. In the worst cases, businesses with long-established RA departments have parts of the team performing routine operational tasks that are normally performed elsewhere in a communication provider. This is not only potentially inefficient; it also raises serious questions over the objectivity of the RA team to assure the operational tasks that they routinely perform.

Revenue Assurance versus Revenue Maximization

Eric Priezkalns

I knew from the very beginning that Ashwin Menon was a very, very, smart guy...

> I have followed your blog since January 2007, and I find that you have touched on a lot of issues/concerns/concepts in the field of revenue assurance. One concept in particular fascinates me—the differentiator, in employing a revenue assurance methodology and a revenue

maximization methodology. I find the lines to be blurred and I would be interested to know your interpretation.

This question arises because in the Revenue Assurance Maturity Model, I see (or I imagine I do) that at rank 5, the communication provider is ideal and moves beyond the traditional viewpoint of RA. For me the logical extension would be revenue maximization. Or is it that revenue assurance and revenue maximization work in parallel?

Thanks and Regards,
Ashwin Menon

I must admit that, before Ashwin asked, it never occurred to me to discuss the relationship between revenue assurance (RA) and revenue maximization. The question is valid, so I was inspired to give a thorough answer about how the two relate to each other.

The TM Forum's *Revenue Assurance Overview* (Dunham 2008), defines RA by the methods it uses—data quality and process improvement, rather than its goals. That definition has been carried forward through all its versions and stems from the first meeting of the TM Forum's RA team, in 2004. It caused a lot of discussion at the time, but what I particularly like about the definition is that it focuses on what RA people say they do when they talk about the specifics of what they do, and not on what RA people say they do when you ask them to define what they do. By focusing on the methods of RA, which are fairly consistent, and the kinds of goals that are satisfied, which are also consistent, it is possible to avoid elongated and unhelpful analysis of whether any specific goal should be adopted by RA. This is beneficial because the general types of goals adopted by RA are vaguely common, being mostly financial in nature, but the specific goals set for RA do vary considerably from communication provider to communication provider. They cannot be generalized while being stated too precisely. I can see why people want to define RA in terms of things like completeness and accuracy of revenues but the problem is that not everybody agrees on those being the goals, or if they agree, they do not do always really aim to meet those goals in practice. For example, completeness of revenues is a very laudable goal, but which executive is going to agree to improve the completeness of revenues from 99.99% to 100% if the final fixes cost more than the 0.01% of the revenue gained? Also, some executives are more interested in ensuring the accounts are correctly stated, others that the bills are right, and others that they get the cash from the customers. The exact measure of success is hence bound to vary, not just from company to company but even within a single company from time to time. This is why I have never been keen on definitions that picked a specific objective as the basis of RA.

The other thing I like about the TM Forum's definition is that it neatly sidestepped one problem raised by some other proposed definitions of RA. Some definitions very simply and elegantly stated that the goal of RA is to maximize revenues.

That is very neat, but very wrong, if you ask me. You maximize revenues by selling and selling, until you can sell no more, but the average RA practitioner is certainly not a salesperson. Indeed, one of the few things that all RA practitioners have in common is that they deliver benefits without trying to make new sales. They are trying to improve revenues, cash flows, or whatever is the chosen measure of success without trying to increase sales. In other words, they improve results without seeking to influence demand or consumption of the communication provider's products. How they do that—well, this is where the uncontroversial core of RA practice comes in—is by trying to identify services rendered but not billed, billed but undercharged, or charged but without the cash being collected, and so forth.

What is revenue maximization, in contrast? Well, if you type the words into Google, you get a very consistent answer back. This answer also accords with my professional training. It is about setting prices to make the most money. Revenue maximization is a basic concept in economics. Businesses may be profit maximizing or revenue maximizing in nature. To get to the maximum, the business tries to work out what products to offer and how much to sell them for. In particular, people draw a lot of graphs, do some equations, collate a lot of data to gauge price elasticity in the market (in short, how much demand for the product goes up or down as the price goes down or up) and where a product is placed relative to competitors. I have no desire to reinvent the definition of revenue maximization just to suit some people working in communication providers, so I would be happy to leave the meaning of revenue maximization as it is—in which case, the relationship between RA and revenue maximization is very clear. One has nothing to do with pricing and demand, the other has everything to do with pricing and demand. What they have in common is data, which in particular may be data about how current customers behave, what they purchase, and how much they pay. RA is looking for flaws or gaps in the data. Revenue maximization trusts the data, but uses it to determine how to influence customer behavior.

Does this mean that revenue maximization is the natural next step for RA, once RA is completely mature? Well, yes and no. Starting with the answer "no," even modest communication providers already employ people who do some kind of job of looking at data and competitors and setting prices. They may just follow the price leaders in the market, but that is still a genuine strategy for revenue maximization. It may not be a very sophisticated strategy, but the point is that this kind of work already predates RA in most communication providers, so it would be misleading to suggest that revenue maximization is a job that everybody has ignored and is just waiting to be done when RA people finally have the spare time to get around to it. Somebody is already doing it, and chances are that the senior executives are very aware of what is being done to enhance revenues. Indeed, in any liberalized telecoms market, the executives are probably far more interested in pricing strategy than they are in RA. On the other hand, the people responsible for revenue maximization may not be very sophisticated at understanding or using data. They may not have all the relevant data. In contrast, RA people may, especially when they have stopped all the leaks, have very good and interesting data, and the skills to use it. They may spot opportunities or pitfalls that the pricing and marketing people miss. For example, they may notice cases where the sales price for a product is less than the cost of supplying it. They may identify network black spots where many customers find calls are dropped or times of a day when congestion stops people from making calls or causes systems to be overrun and calls to be given for free. They may identify segments of customers who demand a lot of customer service time or a lot of credits and who churn after obtaining handset subsidies that are greater than the value of the calls they make. They may identify the signs of arbitrage and of seemingly honest customers who always pay their bills but who are actually connecting VoIP services illegally to the network to avoid termination fees. In summary, the data obtained by RA, when combined with the skills to analyze that data, may yield many benefits for the company in terms of selling more to desirable customers and selling less to undesirable customers.

RA typically deals with issues that have fallen through the cracks. If nobody else deals with the problem, then RA is there to sort it out. That mentality is appropriate for the examples listed in the previous paragraph. All of those problems may have fallen through the cracks. It is in the nature of any organization to be confronted by problems that have not been anticipated and where there is no clear responsibility for who deals with them. That does not mean these specific problems must be allocated to the nominal RA function. It also does not mean that RA people have an automatic mandate to interfere in these cases either. There are other skills than the skills that RA practitioners have. For example, I have often heard RA people talk about the impact of changing price and the effect this will have on customer behavior, most usually with the point of view of exploiting inelastic behavior. For example, a communication provider could increase the price and the customer will buy as much, hence leading to higher total revenues. While it is tempting to make pronouncements like that, RA people may simply lack the data to justify their opinions on matters like these. It is fair to assume that a few pence or cents here and there will not be noticed by customers, but there is a point where any customer may look around for a new supplier when they notice their bills are getting higher although their phone use has not increased. Many judgment calls need to go back to the pricing and marketing people. They can obtain alternate sources of data from market research that typically fall outside of the intelligence made available to the typical RA practitioner. RA can help a lot with using the business' data, but should not be handed freedom to dictate wider business policy as a consequence. I do not think anyone is openly suggesting that is what they mean by the use of the phrase "revenue maximization" when spoken in circles that include RA people, but this takes us full circle back to definition of revenue maximization and even those people who thought RA was the same as maximizing revenues.

In the end, dedicated RA departments have a job to do. That job uses certain specific skills and tools. There is a role to be played. It is possible to step beyond the goals of RA and to use those skills and tools to assist in the realization of other goals that also help the business and fit with its priorities. This may naturally take place as a skilled RA team reaches maturity and finds it has experienced people who are not as well utilized or as fully challenged as they might be. It can also take place sooner. Just because RA is immature does not mean it is always a bad idea to redirect resources to quick wins that fall outside of the remit of RA as defined. Sometimes the value of those wins will make it worth taking a detour from the RA highroad. However, while RA has a clear role that will necessitate some dedicated staff, revenue maximization does not. It is too nebulous to become a new full-time source of employment for people who have reached the end of the RA evolution. At best, revenue maximization is always going to be the responsibility of many people within a business, not least the pricing and marketing functions. There may be a need for some individuals to move from RA to roles that support the revenue maximization goal, but those cases will depend on the history, challenges, opportunities, and current resources in their specific business. It will not be a universal norm that RA turns into revenue maximization. Instead, RA needs to focus on what it does best, as well as the reasons why it is needed. If the skills with data and analysis are there, they can be put to use to assist in areas that are beyond the normal scope of RA. Being helpful, though, is a long way short of taking ongoing responsibility. While it is glorious to see revenue practitioners be ambitious and succeed, it would be unwise to assert that all should aspire to take on the broad responsibility hinted at by the far-reaching phrase "revenue maximization."

Billing Assurance and the Ends

Eric Priezkalns

Some years ago, Vodafone UK, had an unusual attitude toward revenue assurance (RA). They had a Billing Assurance department and they believed that RA was the job of one person in that department. The RA job was responsible for the validity, completeness, and accuracy of all the data that reaches the billing system. Once there, the responsibility handed over to the rest of the Billing Assurance department. That was very unorthodox, even back then. Most people seem to agree that RA is the end-to-end job and that billing assurance is the subset concerned with the billing end of the chain. There is also increasing recognition of the fact that, while symptoms of problems usually occur in billing, the causes are typically earlier in the processing chain. That means you need to spread the RA effort across the entire chain, beginning with the collection of data reflecting new orders, network events, and such.

Where the chain starts and stops, and where the most effort is needed, is not something people agree on as often. When people say they do E2E assurance they often pick a convenient definition of what the ends are. I mean convenient in the sense of what they know how to do or convenient in the sense of mirroring the span of their authority in the business. Not many people doing end-to-end assurance really do much work to ensure the very start of the charging process is correct, which is to say they do not check if all orders are taken correctly or that switches are configured properly. There is even more disagreement about what is the other end. You would have to be very unconventional to exclude billing from the end-to-end chain covered by RA, but there are differences of opinion about how far to go after billing. Some stop at the output from the billing system in terms of the feed to the print vendors for the bill. Some go further and really check whether the bills are correctly printed and sent. Some think the goal is correct reporting to the general ledger. Some think the end point is the collection of cash and the maintaining of customer accounts. So while most people act pretty confident when asked what the scope of RA is, you often get the feeling that there is some fraying at the ends.

We can ask ourselves two things to help frame the question of how billing assurance relates to RA:

- How should we define the role of billing assurance in a company that already has an RA department?
- Where does a billing assurance function fit into the company structure?

Here is how I answer those questions. I have worked with communication providers with a separate Billing Assurance department and with communication providers that did the same work without having a separate department. Fundamentally, I do not think there is one right answer: it depends on the business. Do not be dogmatic. At this level, organization structures are most influenced by the personal and political considerations of the people involved. It might be nice if executives tried to be objective and worked out what was best on a more analytical basis, but they also need to get the best from their employees and that may not be the same as a textbook answer.

After saying that, I will try to give a textbook answer. The consulting/software firm Cartesian has issued and maintained a very helpful "jargon buster" document (Cartesian 2010) where they canvassed opinions and used them to define terms. As a consequence, they formulated different definitions for billing assurance and RA. Cartesian consider the end goal of both RA and billing assurance to be the same, but that billing assurance is the strict subset of RA that relates solely to the billing system(s). That seems like a decent definition to me. Cartesian went on to comment that it is important to distinguish billing assurance within RA because it involves specialist knowledge of how billing systems and the processes around them work. If you accept this line of reasoning, you have a straightforward choice. First, you could place billing assurance in the department with the best specialist knowledge of billing. This might be a Billing department in Finance or it might be a Billing Operations department in IT. The advantage is that you try to have the relevant skills and knowledge in one department, making communication, training, and recruiting easier. The disadvantage is that you have no separation of responsibility.

The Billing Assurance team may not do a thorough job of checking what their colleagues are doing. They may suffer the same misconceptions or be blind to their faults. The second option is to separate billing assurance from the people they oversee, to give some independence. The key here is that the more separate they are in the company structure, the greater the objectivity of billing assurance, but also the greater the overheads in terms of recruiting and training people in billing assurance, and the greater the overheads in terms of communication between billing assurance and the departments that they scrutinize. The third option is to have a blend between the first two. In this case, some detailed day-to-day assurance work is performed by the individuals responsible for the processes and systems, and some higher-level assurance is conducted by a separate function, which provides a form of internal control or audit.

The right choice depends on company culture and dynamics. A business that is entrepreneurial, is smaller, is in a rapid growth phase, or has flat management that should probably keep billing assurance responsibility with the people who do the work that needs to be assured. They sacrifice independence for flexibility. Imposing strong lines of division may otherwise lead to underfunding of billing assurance or too many internal conflicts. A larger, more static, hierarchical, cash cow/cost-conscious business is better off with separating billing assurance from the billing being assured. They can afford to spend more time and money on staff development and training to realize the financial and governance benefits of greater independence.

Let me finish with one final, key, observation. In many communication providers, both the Billing department and the RA department report to the same person. Hence, billing assurance will definitely report to the person who is also responsible for billing. This means there is no independent route to report to executive level about billing and its flaws. This may undermine independence when it is most needed. But having the RA/Billing Assurance department and the Billing department report into different CxOs can be unwieldy. A sensible compromise would be to require Internal Audit or Risk Management to annually review and report on the adequacy of billing assurance to provide some high-level independence.

Conflicts between Revenue Reporting and Revenue Assurance

Eric Priezkalns

Attending a conference, I was lucky to meet Lior Segal of Partner Communications, the Israeli wireless operators who market themselves under the Orange brand. I was lucky because Lior is a CPA, MBA, and attorney-at-law, as well as being chief of his company's Sarbanes–Oxley & Revenue Assurance Department, thus making him the ideal person to challenge with one of my favorite questions: how to balance responsibility for accurate reporting with the pressure to improve the bottom line? I will not try to reproduce from memory the elegant and considered reply he gave, as I doubt I could do it justice, but I will state that Lior did not shy away from the central problem: that there is the risk of conflicts between those goals. Lior instead took the time to explain how he ensured those risks were managed and mitigated. Step one in risk management is, of course, to identify the risks that need to be managed. That is why I have so little patience with consultants who talk about revenue assurance as a panacea for all leakage and governance ills, without highlighting that a seeming cure for leakage may make governance worse, or vice versa.

So while I was lucky to meet Lior, he was rather less lucky to meet me. As well as asking awkward questions, I lack the elegance to jump straight to a good answer. Instead I just try to bulldoze my way through all the options and see which one gets damaged least. The simplistic solution to potential conflicts between reporting and assurance is to have a strong separation of duties. Everyone can agree that is ideal, especially if you want to live in a bureaucratic world where 10 people are employed to do a job in the hope they do it slightly better than one person would have done if left to do it on their own. Separating responsibility for the integrity of reporting from the job of closing revenue leakages is similar to separating the assurance of billing from the people who actually manage billing. Separation is great in an ideal world, because in an ideal world everybody works hard, knows what he or she is doing, and is competent, knowledgeable, and trained. That sounds a bit like a utopian dream because it is; getting lots of people to work hard, and have sufficient knowledge, skills, training, and so forth, usually means spending lots of money. If you can get a similar caliber of job done at much less cost, most businesses should be happy with that.

One complication is that although the knowledge needed for revenue reporting and revenue assurance may have many overlaps, the skills needed may be very dissimilar. For example, revenue reporting does not require imagination, unless you work for someone like Bernie Madoff. Revenue assurance, however, should involve imagination, because of the need to anticipate what might go wrong before it does, or to guess where to employ resources looking for leaks. Without imagination, assurance may end up as useful as health and safety checks on the Titanic. Another complication is motivation. I have seen more than one revenue assurance department become deflated after they were excluded from competitions asking employees to suggest ways to cut costs or boost revenues. The people in revenue assurance spend much of their time inventing and executing similar ideas, but their only reward is their regular pay packet and not much thanks. However, if you give them financial incentives, that can cause problems for reporting, because they have the incentive and detailed inside knowledge to find ways to cheat the system. For a start, they are usually the only people able to report on the benefits they add. In fact, their knowledge may be so detailed and so specialized that nobody else may be able to identify when they cheat the system.

The trick here is to strike some kind of balance: realize cost benefits for educating and developing staff together, wherever possible, but make sure assurance and reporting have different objectives and the people who work at each goal look at the world differently. Try to avoid people assuring the things they also report on, and where it is too costly to avoid, make sure somebody else is checking that everything is fine (and that they have no motivation to cheat either!). By all means allow people to move jobs backward and forward, making for strong cross-fertilization of ideas and know-how, but ensure that the rewards and expectations for each kind of job are distinct and do not get confused. All of these are more easily said than done—which is why I am glad it is Lior's responsibility and not mine. CPA, MBA, attorney... Lior needs to be the lot.

Sarbanes–Oxley + Making Money = Confusion

Eric Priezkalns

Is it possible to confuse accounting integrity and business profitability? I believe so, as evidenced by a lot of bad advice from people who should know better. Ever since people found they could sell advice about Sarbanes–Oxley Act (SOX) compliance and revenue assurance (RA) at the same time, it has occurred to them to try a both-ways sales bet, saying that if they can deliver one, they can deliver the other. But when people start using phrases like corporate governance and RA in the same sentence, be careful, as not everything may be what it seems...

Now, we all know that RA has something to do with revenues. The clue is in the name. For the most part, it is about employing people to do things that improve revenues. I doubt anyone would find that last sentence to be controversial. You may also have read about, heard about, or perhaps you know someone responsible for an accounting scandal. Most scandals are to do with exaggerating revenues. That statement should also be uncontroversial. Put those two statements together and, hopefully, you can see why I am bemused when accounting professionals fall into the trap of saying there is a natural synergy between improving revenues and preventing accounting scandals. Not so. Rather than having a synergy, they suffer from a conflict. This means linking RA to corporate governance is not a suitable route to avoid accounting scandals. You do not need to be an accounting sophisticate to see that services sold on the promise of increasing revenues may be ill suited to stopping the dishonest exaggeration of those revenues. But some accounting professionals and consultants turn a blind eye to this contradiction. They do so either to increase their sales as business advisers or to mitigate some of the risk they bear as auditors by putting extra, but ineffective, burdens on their clients. Either way, they have lost objectivity.

Where do I begin in pulling apart protecting your financial reporting integrity from adding value to the bottom line? Let us start with one important word: revenue.

The word revenue has a different meaning when you are talking to different people. Accountants have a precise definition of the word, the rest of the world a rather less precise definition. Put simply, for an accountant to recognize revenue from a sale, there has to be reasonable confidence that the business will eventually get the cash owed to them. So in accounting scandals, dishonest people pretend that some cash will come to the business, even though they know it will not. Everything comes back to cash, and there are dangers when falling in love with flattering revenue figures. As the saying goes, "revenue is vanity, cash is sanity." The trick for dishonest people is to escape the company before an irregular relationship between revenues and cash inflows is discovered. If some unexpected extra cash does turn up, they may get away with the crime by pushing a few numbers around and hoping nobody notices anything odd. However, if that cash does not turn up, you end up with a scandal, the restatement of accounts and quite possibly insolvency. But most people use the word "revenue" in a much less precise way. They mean something that is only barely more precise than "good stuff that we like." So when people talk about "revenue leakage," they similarly mean "bad stuff we do not like." It is perfectly possible to report accurately the revenues in the accounts while also suffering very large revenue leakages. Revenues are apples and revenue leakages are pears. Linking the accuracy of reporting revenues with the challenge of reducing revenue leakages is bogus. You might as well argue I cannot accurately report how many apples I have if I do not know how many pears I am losing.

In the phrase "revenue leakage," we are talking about what might be called "could have been revenue" or "should have been revenue" or "might have become revenue in the right circumstances." For the accountant, it is only revenue when you can be confident it will eventually turn into cash. Confidence is important. Accountants are prudent. If they are worried that the cash will not come in, then they cannot count the revenue. So fixing a revenue leakage may never lead to revenues. Take a simple example: imagine some sales data gets lost underneath the billing system or down the back of the sofa. The RA man looks around and finds it a year later. Hoorah! A leakage spotted and a pat on the back for him. Now all we need to do is to raise a back bill. Can we be sure we will get the money and hence claim the increase in revenues immediately? No. Perhaps we discovered the customer was a fraudster and we have no way of sending them a bill, never mind any hope of having it paid, or the sales refer to lines that the customer did not know he or she had because the communication provider failed to keep and communicate proper inventory records, and the customer has since ordered and paid for alternate lines. The latter customer cannot be expected to pay twice because the communication provider's goof up caused it to supply effectively the same service twice. I could go on with further examples of why the backbill should not be raised and would not be paid even if it were raised. There are 101 reasons why fixing the "revenue leakage" may not lead to extra revenues.

The important point here is prudence. When people say they lose 2%, 5%, 20% or 50% of their revenues in leakage, they are doing no such thing. They are losing exactly 0.0% of their revenues. You cannot lose something you never had in the first place. What they are losing is "could have been/should have been/might have been revenues," which is quite different. A revenue leakage is no different to a salesman who is rude to a customer and so loses a potential sale. If they had been a bit nicer they would have gotten the sale, and the revenues would have been higher, but that is a big "if." The truth is they were not nice and so they did not get the sale. Similarly, with revenue leakages, if you had fixed some bug in some system you might have raised a bigger bill. But if the communication provider did not fix the bug, it did not raise a bigger bill and it did not get the revenue. Meanwhile, in the accounting world, the revenue figure reported is perfectly accurate irrespective of leakage, because it is supposed to exclude all those could have beens, maybes, or ifs. So it is a confusion to suggest you need to fix revenue leakages to get accurate revenue reporting. Fixing revenue leakages is just about running a business well, in the same way as training staff not to be rude to customers is about running a business well. Categorizing it as a governance issue gives it a mystique, but it is no more a governance issue than offering good customer service or motivating employees.

The next, more serious observation about linking the reduction of revenue leakage to preventing internal accounting fraud is that they are distinct risks. That makes them both part of risk management, but the nature of risk management is to respond to the actual risk. Once again, the clue is in the title. Crucially, risks tend to have a direction. This is because people are inclined to do naughty things one way, but not the other. An employee secretly stealing company property is a risk. An employee secretly making a charitable donation to the company is not a risk. Accurate financial reporting is about telling the shareholders how much revenue has been earned. The risk is that the revenues are exaggerated because executives, managers, and employees may be rewarded for good results and fired for bad results (I can think of situations where executives might want to understate revenues, but those situations are the exception to the rule). So good risk management will generally focus on controls to prevent overstatement of revenues. However, the justification for RA is that it delivers higher revenues. Taking the example of businesses with inadequate controls to prevent overstatement of revenues, we end up with an argument for spending time and effort on activities to increase those same revenues. Admittedly, the controls to achieve one are sometimes the same as the controls to achieve the other. However, often the controls needed for each objective are very different. We should not assume that both goals can be realized using the same control, simply because that would make life more convenient.

From a governance perspective, shareholders want the revenues reported in the accounts to be no more maximized than they want them minimized. What they want first and foremost is accuracy and reliability of the information given to them. The risks of inaccurate financial reporting are distinct from operational risks related to poor data integrity. One is about the deliberate intention to deceive. The other is about data processing errors that nobody is aware of or responsible for correcting. Safeguards to prevent dishonesty have little to do with safeguards over processing.

Imagine you are an executive (or if you are an executive, just read on). What is the risk if you understate revenues? You wake up one morning and discover that because of some screw up the company has more cash in the bank than expected. You quickly work out where all the extra cash came from, or at least dream up a plausible story. You tell the shareholders, and they seem very happy. The sun shines on the nice new car you buy to celebrate.

What is the risk if you overstate revenues? You wake up one morning and discover you are Bernie Ebbers, former boss of Worldcom. You can ask for parole in 2028, when you will be 85 years old. You see the sun shining through the bars of your prison cell.

Being lousy at running a business is not a crime. Lying to the owners of the business by making up bogus numbers is unlawful. RA helps executives run better businesses. Corporate governance scandals are about a failure of honesty. Linking the two by saying there is a simple win-win relationship between RA and corporate governance is wrong. Whenever CPs turn to consultants for help, they must be sure they draw a line between talking to someone as an accountant and talking to someone as a business adviser. The accountant should be mindful of the responsibilities to shareholders, who need accurate reporting so they make informed investment decisions. The business adviser enhances business performance. These interests are not harmonious. That is why we have financial audits, and hence why we have accounting firms. Excessive attention on increasing apparent shareholder returns and insufficient concern for controls over reporting leads to dishonesty. Trying too hard to devise solutions to improve performance and increase control at the same time will put one or both of those goals in jeopardy.

To be fair, the accountant-consultants also talk about the risks of overcharging, and that does have a more natural (though still imperfect) fit with SOX compliance. But that accentuates the point about not linking financial integrity with financial gain. Why talk about maximizing revenues, while also admitting that you need to implement RA controls to prevent overcharging? Fixing overcharging could just as well be described as minimizing revenues. Why set a biased expectation that revenue integrity will lead to increased revenues when there is an acknowledged possibility that revenues may fall as a result of correcting overcharging errors previously missed by both the business and its customers? We can also consider fraud and its link to SOX compliance. But again, the frauds described as ways to improve the bottom line have nothing to do with financial reporting scandals. There is a big difference between being ripped off by a fake customer and someone inside the business massaging the results to secure their bonus. The kinds of frauds described as "leaks" tend to be crimes where no additional revenue had been recognized by the business and where the fraudsters would never have intended to pay for the services acquired anyhow.

"There are three kinds of lies: lies, damned lies, and statistics," said Disraeli. "Ninety percent of statistics are made up" goes an old joke. In one article I read, a Big 4 director described estimates of communication provider revenue leakage between 2% and 10% as "research." Ahem. What were we talking about, again? Risk, bias, misreporting, and that kind of thing. So what is this research he refers to? Surveys of estimates made by people who get paid to fix revenue loss. No risk of bias there then.

Let me indulge in one final anecdote, drawn from a conference I attended back in 2005. There was a panel discussion of the linkages between RA and Sarbanes–Oxley, hosted by a Big 4 accounting firm and including a panel of three leading personalities from telco RA. After a half hour of listening to the Big 4 partner spiel on about how well RA and SOX complemented each other, with not a single mention of the possible conflicts between the two, you can imagine how keen I was to throw a spanner in the works. Luckily, I got my chance when I asked the panel to answer, in a word, whether complying with SOX now meant they had eliminated all material revenue leakages. To their credit, all three of the telco panelists—Lionel Dawson of O2, Patrick Halbach of Qwest, and Moly McMillan of NTL, answered the question frankly and with a resolute "no!" By doing so, they begged the question of how much advantage there really is in trying to link the objectives of RA to those of good corporate governance. The most useful synergy is in marketing for the Big 4—if they cannot sell their services by promising it will deliver compliance with obligations, they sell it by promising extra revenues.

It is folly to pretend that SOX compliance and RA is the same thing, even if the folly sometimes proves profitable. There are as many conflicts as synergies between RA and SOX. Motivating people to increase revenues is good for business, but the essential problem is that some people also become motivated to cheat. Programs to install safeguards over accurate reporting cannot start out on the pretext that higher revenues will follow as a natural consequence. And programs to increase revenues will not eliminate the worst kinds of misreporting. Shame on any professional accountant who does not make that clear.

Macrodiagnosis versus Atomic Checks

Ashwin Menon

I will tend, in the course of my writing, to try to displace the accepted norms of RA. The primary driving force behind this desire would be to challenge the way things are currently done so that we can analyze and come up with better methods. The following is a discussion that Eric and I were debating earlier.

One common thing I have noticed in RA implementations is that the communication provider tends to gravitate toward accuracy of transference (xDRs from one network element being recorded in the corresponding downstream system) as the crux of RA. In my opinion, transference is simply a symptom rather than the disease itself (e.g., a voice CDR produced by the mobile switching center (MSC) not being present in the downstream mediation platform might be because of incorrect provisioning of a postpaid customer, where the CDR stamping at the MSC defines the call as a prepaid call). However, in accordance with the drip-tray model (see Dunham 2008) I can see the validity of performing transference checks, but I feel that RA as a discipline should begin to recognize the benefits of performing atomic level checks instead of the macrodiagnosis model that is currently prevalent.

What am I suggesting here? Simply put, check the health of the underlying system before we check the output of the same. A simple example would be rating. In my opinion, there are two ways to perform checks on the rating engine. One, we can rerate some sample xDRs and check whether the system rating is in line. Second, we could instead perform a reconciliation on the underlying rating tables, which forms a critical part of the rating engine. As anyone related to a rating function in a communication provider would tell you, maintaining a separate parallel rating framework to rerate xDRs would be a massive task. It would be so much simpler (and more cost efficient) to validate the rating structure itself within the rating engine.

Macrodiagnosis does have its benefits (as summaries help to get a bird's eye view of the overall health), but I believe in the intrinsic value of performing atomic level checks as they would be more cost efficient as well as being beneficial in terms of root cause analysis.

The Zoom

Eric Priezkalns

Do you ever think revenue assurance (RA) is too broad? Do you sometimes feel that the list of expectations never ends? Or do you know people who work in RA who, in one sense, do a reasonable job, but somehow miss the point?

The article by Ashwin about macrodiagnosis and atomic checks links to the same themes that have inspired Güera and her academic research. That is very revealing. We are talking about two RA experts from very different backgrounds—a woman in South Africa with a background in psychology whose experience comes from Africa and North America, and a man in India who is an engineer with most experience of communication providers (CPs) in the Asia-Pacific region—yet they are talking about the same phenomenon, one which I believe is unique to RA. I have my own pet name for the phenomenon: I call it the zoom, and it makes good RA quite unlike most other jobs.

You know about zoom lenses on cameras. You can set the lens so the camera sees everything from a distance. Twist the zoom, and suddenly it is focused on a tiny piece of detail. For similar reasons to why cameras have zoom lenses, RA practitioners need their own kind of zoom. Their zoom works with their mind, not with their eyes. They need to be able to stand back, and see the big picture. They need to do this to have a clear understanding of objectives, and to understand how a root cause in one part of the business leads to undesirable symptoms identified elsewhere in the business. Practitioners also need to be able to dive into the detail. They need to be able to analyze individual data records or understand specific lines of code. The zoom is one of the most important skills in RA, yet it does not have a proper name. Somebody with the ability to zoom in and out, who can comfortably shift between the micro and macro and see how they are connected, will be better at RA than someone who cannot. Yet you never see the zoom being properly described in job profiles. You may see job profiles that ask only for certain levels of analysis— sometimes micro only, sometimes macro only. You may see job profiles that ask for an unrealistic level of micro experience across too many fields, but fail to mention the real skill that is needed: the ability to learn, understand, and analyze the detail as and when necessary, which is not the same as knowing it all already. You may see job profiles that ask for big picture thinking, and links this to an arbitrary subset of detailed OSS/BSS knowledge; their prejudice about what subsets of detailed knowledge are needed only undermines the real requirement to have an open mind about where the root cause of problems may really lie. What you really should see is a job profile that asks for the zoom.

I doubt we ever will see a job profile that asks for the zoom, and not just because people may find the phrase funny and so not take it seriously. In CPs, some people (especially engineers) get paid a lot of money to know the detail. Other people (especially executives) get paid a lot of money to see the big picture. They are supposed to work together and deliver services that please customers and processes that satisfy the needs of the business. However, the job is too complex and they make mistakes. That is where RA comes in. This is why RA people need to be able to work with detail and the big picture, to ensure everything is correctly aligned. But hang on .... If people get paid a lot for detail, and people get paid a lot for the big picture, surely that means RA people need to get paid the engineer's salary plus the executive's salary? That is silly, of course. You are not going to solve the problems of big complicated companies by employing people to act like all-knowing gods, aware of everything that is going on at every level of detail and abstraction simultaneously.

Tempting though it is to try to solve the problems of CPs by giving somebody the job of solving those problems, they will not succeed if the task is not humanly possible. If the solution was to train an RA person to know everything, at every level from macro to micro, you might as well train the people already in the company and save the cost of paying someone to do RA. The real solution is to employ an RA person who is skilled in communicating and working with people, and can use the skill to link the detail to the big picture, and highlight the inconsistencies. But that skill sounds a bit abstract, so employers get scared and ask for kinds of knowledge for which it will be easier to test if the person already has it. They ask for practical knowledge like the ability to write SQL queries, knowing the components of a GSM network, or knowing the components of the COSO ERM framework. This kind of knowledge is easier to check in an interview, but not relevant—knowing how things worked in the past is not a reliable indication of the aptitude to learn about things in future, or knowing some detail does not mean the candidate knows the detail that will be most pertinent to the issues that are actually plaguing the CP's business. The point is not about knowing the intricacies of, say, GSM networks, but in knowing about how to obtain the key information needed for RA, and knowing how to apply that information once you have it. If you can do that for one kind of network, you can do it for any other. The same applies to business goals. One business may want to increase margins, another wants to increase customer numbers. A good RA practitioner will understand how to prioritize his or her work to reflect best the priorities of the business, which change over time anyway.

The zoom is about managing complexity. The skills needed to manage complexity should be what makes RA a unique practice. It is not RA's job to know the detail better than another part of the business, or to tell executives what the big picture should look like. RA adds value if it acts as a bridge, highlighting where the detail and the big picture are not aligned. Of course, people do that all the time as part of their jobs. What makes RA special is that it addresses the misalignments that are not obvious to anyone, because they are lost in complexity. To manage complexity, you do not start by trying to know everything. Knowing everything just means copying complexity, not managing it. To manage complexity, you need to be able to break it down, identify the essentials, and ignore the nonessentials. This draws upon a variety of skills, like being able to identify patterns in large volumes of data, or being a careful listener who can tell if person A said something inconsistent to person B. People who lack the zoom will inevitably look for root causes in the wrong places, or work counter to the real priorities of the business, or advocate solutions to problems that are far less efficient or far more costly than alternatives they failed to consider. The zoom is the essence of good RA. The reason why we need RA is that most people cannot or do not zoom from big picture to fine detail as part of their normal job. The failure to zoom is the blind spot for most businesses. That is part of the reason why we lack a good name for it. We need the zoom; the challenge is in seeing that we need it.

Being Proactive

Eric Priezkalns

Here is a new thought. I have seen quite a few scenarios where an operator claims that he has a proactive check system in place, but on further analysis, we find that the checks and the system are not truly proactive, but merely reactive with a much smaller delay between the timing of when something goes wrong and when it is discovered. How can a communication provider systematically set about becoming more proactive in their execution of revenue assurance (RA)?

This question got me thinking about the nature of revenue leakage. When we take a step back and look at the big picture, we see that a leakage happens due to unforeseen eventualities, misconfigurations/omissions, poor system integrations, or maybe even internal fraud. In none of the aforementioned cases (which are but a few of the reasons for revenue leakage) can we truly say that we are capturing all leakages. How would we go about proactively checking for integrity and completeness?

In my experience, I find that to a certain level, we can perform proactive checks as far as subscription data is concerned. Standard checks like matching subscriber feature information between the home location register (HLR), provisioning and billing, and so on, would, in some ways, prevent revenue leakages proactively (so long as there is integrity of subscription data, many potential issues with usage data are also avoided). Issues in usage data like usage not being billed can be attributed to unsynchronized subscription information across the provisioning system and billing system. I have seen cases where a subscriber is provisioned for certain services at the switch but the same set of services is missing in the billing system. What happens in such a case is that the subscriber uses certain services (such as voicemail), but will not have to pay anything for the service.

For usage data (i.e., xDRs), I find that most proactive checks still require at least a minor amount of leakage to occur before an alarm is triggered. The only true way to perform proactive RA as far as usage data is concerned is to either ensure all root cause possibilities are covered and alarms are set up at the core level rather than at the data output level (which is a massive task), or have an RA test bed before launching any service into a production environment (we could scale down the total load to consider). Using the test bed approach, we could effectively run RA metrics to verify absence of leakages, as well as capture issues at an early stage. Once the RA department clears the new service/product, it could be launched into production. Once again, this would not give us 100% confidence that there are no revenue leakages, but it would proactively eliminate the possibility of some leakages that normally get identified only after launch.

Chapter 5

Epistemology and Metrics

Epistemology, Icebergs, and Titanics

Eric Priezkalns

When writing a book about revenue assurance, it was evident that the biggest challenge would be unpacking and addressing the often unspoken challenge that defines revenue assurance: methodically working through what we know, and what we do not know, about the errors and problems that cause leakages in communication providers. If we dive for detail too soon, we focus on what we know, and may miss the subtler trick of trying to identify what we do not know, but that we should know and we need to know to get a complete picture. Epistemology is the branch of philosophy concerned with the nature and limits of knowledge. Much of the writing on *talkRA* has revolved around the epistemology of revenue assurance, and it was inevitable that a chapter of this book would be devoted to this elusive and often misunderstood topic.

Consider the following: you are a mechanic and it is your job to inspect a motor vehicle, to determine the condition it is in. To make your life easier, I provide you with a checklist. The checklist details a number of the car's parts and systems and what tests you should do to determine if the parts and systems are working correctly or are in need of maintenance. You might accept the checklist on trust, but how do you know the checklist is complete? What if there was another system that should be covered during the inspection? What if this system was faulty and the fault has serious implications for the driver and passengers of the car? If the responsibility for the completeness of the checklist is yours, and not the responsibility of the person who gave the checklist to you, what should you do to ensure the checklist is complete? Just looking at the car might provide some pointers, but could never eradicate the doubt that there was something you did not think or know to look for. Now, take this thinking and apply it to revenue assurance. How, in short, do you determine the limits of what should be checked, to prevent the business from suffering leakage? At least with a car you could literally take it to pieces, and separately identify every component, thus knowing if you had checked every one of them. When dealing with something as intangible as the data and processes of a communication provider, how do you do something similar?

I will not provide an answer to the above question here, not least because the following articles discuss several techniques to build a complete view of the task faced by revenue assurance. None of these techniques are one-stop-shops that arrive at a definitive and final answer. All are contingent; to get the best results, you keep having to ask the question of whether you have missed something, and keep exploring for new kinds of leakage not identified before. What I will do in this article is to introduce two metaphors that I have found helpful for explaining the reason why the challenge is so important. Simple analogies can often be vital to expressing why work needs to be done to find a problem that nobody has found evidence of before. Without the buy-in to expend resources on challenging the limits of your knowledge, you may have a comfortable working life, while everyone remains ignorant of the real leakages that hurt the bottom line because nobody knows they exist.

The Leakage Iceberg

Figure 5.1 illustrates something that we all know about icebergs—only part of an iceberg is visible from the surface. If we think of all leakage suffered by the communication provider like the mass of an iceberg, only some of the total is visible to us—visible in the sense of being measured by the monitoring activities and checks of revenue assurance—and some is invisible because we have no data. This is a very

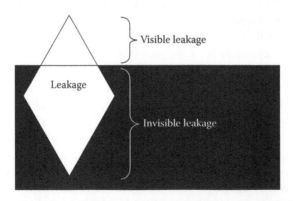

Figure 5.1   The leakage iceberg.

real problem. Indeed, the reason why we need revenue assurance is that leakages can go unidentified unless a focused effort is made to identify and contain leakages. The challenge is for revenue assurance to make all leakage visible. Improving revenue assurance controls means raising more of the iceberg above the water line, bringing it into view so we can see the true scale of leakage being suffered. Once we have visibility of leakage, we can then follow a methodical and prioritized way to reduce it.

The Titanic Effect

The book *Computer Ethics* (Forester and Morrison 1994) gives a detailed analysis of why errors may go undetected in computerized systems, sometimes with fatal consequences. The authors identify a vital principle, and they do so by drawing on an analogy that predates the computer era. To their mind, one factor that contributed to the sinking of the Titanic was overconfidence in the design of the "unsinkable" ship. The Titanic effect, which also applies to automated systems, is that the more confident the designer is that a system cannot fail, the more severe the effect will be when it does.

Conclusion

The iceberg reminds us that the true scale of a problem may not be visible. Some of it may be hidden. We can add benefits, while focusing only on what we already know about, but the consequence is that more severe leakage problems will go untreated. In contrast, the Titanic effect reminds us that overconfidence is a root cause of severe failure. Caution is a more effective long-term strategy than taking a risk that the current design of systems is sufficient to ensure nothing can go wrong. To continue the analogies a little further, it is also well documented what happens when the Titanic comes into contact with an iceberg: leakage on a disastrous scale.

Two Dimensions of Leakage Knowledge

Eric Priezkalns

In Figure 5.2 we see the two dimensions of leakage information. The vertical axis is the one that most businesses tend to focus on, the reported value of leaks, from low to high. The horizontal axis is equally important, though less commonly used. It represents the completeness or coverage of leakage information, from low on the left to high on the right. A business with few revenue assurance controls and measures may report low levels of leakage, but may do so in ignorance of the true extent of leakage. In contrast, a business that reports high levels of leakage may deserve

Figure 5.2 Leakage knowledge dimensions.

praise if the value reported has risen because of the implementation of an increased span of leakage controls. For reporting to be fair to all businesses, both dimensions of information need to be assessed on a regular basis.

We can crudely characterize the four quadrants of business knowledge of leakage as reflective of the communication provider's attitude to revenue assurance. A low level of controls combined with low values of reported leakage represent a business with a complacent attitude to revenue assurance. There may not be a problem, but the business, in general, assumes there are no problems rather than assuring itself of this. A low level of controls combined with high values of reported leakage represent a business in fear of revenue leakage. The degree of knowledge may be inadequate, but the assumption is that leaks are occurring. In this case, nobody can be sure where the leaks are, potentially leading to effort being wasted on plugging phantom leaks. A high level of controls coverage combined with high values of leakage reported is a business that has awoken to the true challenge of revenue assurance. It has the knowledge of where leaks occur and how serious they are. This business knows how to move forward by resolving existing leaks and preventing future ones. A business with a full complement of controls and low reported leakage is in a state of revenue assurance delivery. It is realizing the maximum benefits of revenue assurance by identifying all leaks and succeeding in addressing them and preventing them from recurring. It is possible to trace a typical evolutionary path where most businesses start out complacent, are motivated to invest more in revenue assurance as they enter a state of fear, discover the extent of the challenge as a result of the investment in revenue assurance staff and technology, and work hard to resolve these, leading to delivery. The typical cycle is characterized in Figure 5.3.

If not careful, the business may complete the cycle and move from delivery back to a complacent attitude. This occurs if controls and measures are not maintained and updated, especially in respect of changes to technology and the launch of new products.

Figure 5.3   The leakage dimensions cycle.

How Cable & Wireless Built a Reporting Framework around a Checklist of Leakage Points

Eric Priezkalns and Lee Scargall

This article discusses how to make use of a leakage point checklist to structure revenue assurance activities. It is based on the practical experiences of how Cable & Wireless (C&W) International applied the generic checklist of the TM Forum to its group operations.

The Checklist Itself

The purpose of a checklist is to ensure nothing gets missed simply because nobody thought that a check was needed. A leakage point checklist is hence a list of every imaginable leakage. The aim of revenue assurance is to put in place controls to address all possible leakages.

C&W International developed its leakage point checklist in collaboration with the TM Forum, largely under the supervision of Stephen Tebbett, while he worked for C&W. The advantage of collaborating with other communication providers is that there are likely to be fewer omissions from the checklist as a result. Experienced revenue assurance consultant Geoff Ibbett has since significantly developed the TM Forum checklist, but the principles of use remain the same.

Cable & Wireless Case Study

When rolling out a global revenue assurance program across all 33 operating units in its international business, C&W was faced with an enormous challenge: how to construct a reporting framework that would consistently capture the scale and significance of every leakage that occurred in every different kind of business model, while also identifying the gaps in knowledge. The solution was to develop an inventory of leakage points, and then to adapt to the needs of their operating companies and then support it with automated reporting. C&W was able to deploy a fair and comprehensive reporting framework rapidly by constructing both tools and processes based around a comprehensive leakage catalog that could be applied to a wide variety of communication providers.

The Context

C&W International has 33 distinct operating businesses providing a wide variety of fixed, mobile, Internet, and data services to retail and enterprise sectors. Each unit tailors its product offerings to suit the local market in which it competes. Operating businesses are spread across the Caribbean and Americas, Europe, and Asia. The business cultures and leakage issues faced vary greatly. In this context, it was necessary to improve corporate governance and financial returns by implementing a systematic and consistent mechanism for assessing and measuring revenue leakage in each unit. The challenge was to find a consistent model for reporting and prioritizing leakage issues.

The problem for C&W was to implement a reporting approach that correctly motivated all units to improve the completeness of reporting and then to accurately report the values of leakages identified. The benefits gained from implementing common revenue assurance solutions across many units needed to be balanced fairly against the value to be realized from high-impact problems that were unique to specific countries. Consistent reporting was vital for prioritizing the allocation of resources and for reducing the danger of bias in reported results, whether a bias toward over or under reporting the quantity of leakage, or a bias toward under- or over-reporting the true extent of reporting coverage.

To implement group-wide reporting, it was necessary to have a comprehensive model that would track and categorize all kinds of leakages. This model would need to adequately cover leakages that might be possible in the operating unit of just one country. It was also necessary to have a fair mechanism to highlight the different levels of leakage knowledge in each country. Reporting low levels of leakage through ignorance should not be treated the same as reporting low levels of leakage thanks to a comprehensive control environment. In short, C&W needed a way of aggregating both dimensions of revenue assurance leakage knowledge, coverage, and quantification, and presenting them consistently to executive management.

Groundwork for a Solution

C&W realized it needed a consistent universal catalogue of all the leaks that can occur throughout communication providers. This catalogue needed to be truly comprehensive, addressing all kinds of products, all kinds of cultures and all kinds of technology. To formulate its checklist, C&W adopted a three-pronged strategy:

1. Canvassing opinion and gathering information on currently known and suspected leaks around the C&W group
2. Informally canvassing opinion from fellow operators
3. Collaboration with other communication providers through the TM Forum

The Full Solution

After formulating the checklist, C&W needed to develop a mechanism to gather and compile monthly information on both the coverage of controls and quantity of leaks. To do this across 33 business units and literally hundreds of potential leakage points required an automated solution. C&W engaged vendor Symbox, with a view to bespeaking their incident management and reporting tool for revenue leakage reporting across its international operations. The resulting tool was called Pandora, a system that, like Pandora's Box, captured all the evils of revenue leakage across C&W's international operations.

Alongside the automated system of Pandora, it was necessary to train staff in each country on the principles that underpinned the reporting matrix and on how to interact with Pandora to support monthly reporting procedures and deadlines. The reporting process combined delegated responsibility for populating data with central responsibility for monitoring and validation, resulting in a truly comprehensive analysis. Monthly reports to executives included a breakdown of known leakage and recovery, plus a summary of control coverage.

Results

The analysis compiled via Pandora supports both executive decision-making and corporate governance objectives. As a result, both opco and group executives received previously unimaginable data on the strengths and weaknesses across the group, and a meaningful framework to contextualize relative performance and priorities. Providing group and opco executives with a breakdown, including quantified leaks and coverage gaps helps to balance executive attention between those issues that are "known knowns" where a resolution needs to be implemented, and the areas that are "known unknowns," where work needs to be done to improve visibility and mitigate the risk that leaks may be occurring without management knowledge. Furthermore, the existence of the checklist provides motivation and a context to explain the need to keep on searching for "unknown unknowns," meaning leakages that have the potential to hurt the business but which have not yet been identified as a risk and not yet included in the checklist.

Since group leakage reporting was implemented, C&W International has enjoyed the following benefits:

- There has been a significant improvement in controls coverage across all units worldwide.
- Levels of reported leakage doubled overall, meaning previously unidentified problems were monitored and addressed.
- There was a commensurate increase in the revenues recovered for the group, as new recoveries were made that were previously not considered necessary.
- Benefits for individual operating units are proportionate to where they were on the complacency-fear-challenge-delivery cycle when the exercise begun, with those operating units that have moved through all four stages seeing the most dramatic improvements, but those units that were already more mature at the outset receiving due recognition.
- The grid of measures provides a meaningful benchmark of performance across the group, spurring improvement and cross-learning.

In conclusion, C&W implemented an intelligible but sophisticated conceptual framework for reporting revenue leaks, addressing quantified known leaks and gaps in knowledge. This framework underpinned a worldwide reporting structure. The reporting structure was automated and deployed across many operating units with differing goals and challenges. Irrespective of the disparities in the technology, services, and culture of those units, all have seen an improvement in their monitoring coverage and in their leakage recovery. This shows that a single reporting model can be made to work if it begins with a comprehensive inventory of revenue leakage points.

Metrics for Leakages and Benefits: Making the Key FIT

David Stuart

The number one expectation nowadays for a revenue assurance (RA) department is the ability to quantify its benefit to the overall business. Many of us will have had meetings with our CFOs where targets are set based on the Earnings Before Interest, Taxation, Depreciation, and Amortization (EBITDA) impact. In the short term, this can be great for RA managers as leakages are aplenty and, therefore, the budget to increase our size and scope is freely available. However, beyond the first few years, targets can become more and more difficult to achieve. Executive perception then changes with regard to the "benefit" of the function and headcount and budget reductions soon follow.

What we, therefore, have to do is identify a way of measuring the benefits of an RA function throughout the maturity lifecycle, thus showing continuous benefit to the business beyond the early years. The following methodology aims to do just that. It will provide a standard methodology for quantifying leakage and recovery that does not discriminate against an individual operator on the basis of his or her RA maturity level or core revenue streams—that is, prepaid and postpaid.

Within a communication provider there are an infinite number of ways in which leakages can occur. Therefore, to determine a standard methodology for calculation, we need to define some common variables against which any leakage can be categorized. This categorization will then make it simpler to determine standard recovery quantification, thus allowing for ongoing quantification of business benefits. The following paragraphs will describe the variables, the categories of leakage, and their associated recovery calculations.

FIT Categorization of Revenue Leakage

There are three variables against which all leakages can be categorized that will determine the type and quantification of any recovery:

1. Frequency of leakage: The frequency of a leakage ultimately relates to the root cause of why the leakage occurs. It therefore impacts how and what can be recovered:
   - Recursive leakage is revenue loss that occurs on an ongoing basis, for example, reference data issues.
   - One-off leakage is revenue loss that occurs once for a single event, for example, a system shut down.
2. Identification of leakage (types of RA and FM): When in the product lifecycle the issue is identified will again impact how we can quantify any leakage and recovery. The timing of the identification will be referenced by the RA maturity levels:
   - Proactive identification of leakage occurs before commercial launch, so this is a preventative measure.
   - Reactive/active identification of leakage occurs post the leakage event (or real-time as it is happening).
3. Timing (types of charging): The final variable we need to consider when trying to categorize revenue leakage is the timing of the charging as this will ultimately impact whether an actual recovery can take place:
   - Real-time charging occurs at the same time as the event is occurring and is classically known as the prepaid customer segment.
   - Post event charging occurs after the chargeable event has occurred and is classically known as the postpaid customer segment; however, in this instance it refers to all post event charging, including interconnect partner settlement.

Quantifying Leakage

Now that we have defined the variables that make up a leakage category, we can determine a straightforward methodology for quantifying the leakage. This can be achieved by looking at just one of the variables—frequency.

- One-off leakages: As the name suggests, a one-off leakage is only going to happen once; therefore, there is no need to quantify it retrospectively, nor is there a need to forecast its impacts on the future. Therefore, the quantification value quite simply has to be the gross value of the (actual) leakage event that occurred.
- Recursive leakages: A recursive leakage is the opposite of a one-off leakage and, hence, requires a slightly more complex solution. A recursive leakage is ongoing in the future as well as possibly existing recursively in the past; therefore, what is required is a standard period for which we calculate the issue. The way we calculate leakage will also impact the way we calculate recovery and as most RA teams are judged on the basis of their EBITDA impact on the business, we have to identify a solution that does not encourage a delayed response to a leakage event; that is, if an RA team spots a recursive leakage occurring within 24 hours of its initiation, our method of calculation should not encourage a failure to report the leakage so as to increase its value and potential EBITDA impact. The solution, therefore, is to calculate all recursive leakage for a set period of time. Thus if it is identified quickly, the leakage value/risk remains the same. The recommended period is 1 year.

Recovery Types

We have now seen how leakage should be categorized so we now have to look at the types of recovery that can occur. There are fundamentally two types of recovery:

1. Recovered revenue represents the actual lost money that has subsequently been collected/charged.
2. Averted revenue recovery represents revenue that would have been lost had the associated risk not been mitigated by a resolution action.

In special cases a single recovery can consist of both a recovered and an averted revenue loss; we call this a combination recovery.

Mapping Recovery Types to Leakage Scenarios

This matrix details the type of recovery that can occur for each type of leakage situation along with the period it should be calculated for.

Using Table 5.1 we can now instantly categorize a leakage as one of eight types, each of which maps to its own specific recovery calculation; that is, if a prepaid tariff had been setup incorrectly charging $0.01 per minute instead of $0.10 per minute and it was identified by the RA team 3 months after implementation, the leakage would be categorized as a recursive real-time leakage that was reactively identified, and as there is no way of back-billing the subscribers the recovery is calculated as averted recovery (future losses that have been mitigated).

By putting this all into practice we come up with Figure 5.4 to illustrate the leakage and recovery calculations.

Calculating Recursive Leakage and Recovery (Over a 12-Month Reference Period)

- Day 0 represents the 1st day the leakage started occurring.
- Day X represents the number of the day, after day 0, when the leakage issue is fixed and resolved.

Table 5.1 Classifying Leakage Recovery

		Identification	
		Proactive	**Reactive/Active**
Frequency and timing	**Recursive—Real time**	Averted (12-months potential)	Averted (12 months)
	Recursive—Post event	Averted (12-months potential)	Combination (12 months)
	One-off—Real-time	Averted (Potential)	Recovered (Actual)*
	One-off—Post event	Averted (Potential)	Recovered (Actual)

* Backbill of prepaid customers is highly unlikely; however, if technically possible, the quantification should be based on actual recovery.

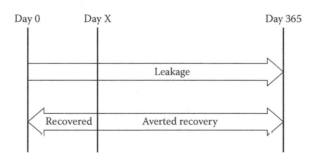

Figure 5.4 Timing of benefit measurement.

- All leakage is calculated over a 365-day period.
- All averted recovery is calculated for a period of 365 − X days.
- All recovered revenue is calculated for a period of X days.
- Note that in a case where the leakage has been identified and resolved proactively, day X is day 0; therefore, averted recovery is calculated for 365 days.

Calculating One-Off Leakage and Recovery (Potential)

Where a one-off leakage is detected proactively (before the event) the leakage should be calculated as revenue loss that would have occurred had the event happened. The associated recovery for this situation should be calculated as 100% of the potential leakage.

Calculating One-Off Leakage and Recovery (Actual)

Where a one-off leakage is detected reactively/actively (post event) the leakage should be calculated as the actual revenues that were lost. The associated recovery for this situation should be calculated as the actual monies recovered, so the value is between 0 and 100% of the leakage value.

Special Cases—Old Leakages

Where a leakage has been occurring for over 365 days before it is identified, the model defined earlier still stands. Therefore, leakage will still be recorded for a period of 365 days; however, NO recovery will be recorded even if it has occurred.

Summary

Utilizing the aforementioned model enables an RA function to track both the actual recovered revenues and the "rescued" revenues through risk mitigation. As an RA function matures and acts more proactively, the percentage of recovered revenue increases relative to leakage. Therefore, the aforementioned model promotes the need for an RA function to follow the progression through the maturity model, rather than stagnate within the reactive/active stages of the lifecycle.

A Strategy for Measuring Charging Integrity Based on the Three "C"s

Eric Priezkalns

This article will provide a generic framework for calculating the integrity of charges presented to customers. The aim is to provide a consistent method that can be applied to any business, and can keep being consistently applied even while systems and processes change.

What Charging Integrity Means

All businesses need to have a means of recording sales and then charging customers for those sales. In some cases, the two are simultaneous. In other cases, the exchange is asynchronous; the customer may execute a monetary transfer before (prepaid) or after (postpaid) the desired goods or services are provided. Whatever the convention applied, any reputable business will want to

- ■ Correctly record what was supplied to customers
- ■ Correctly charge customers for what was recorded as supplied

The first bullet relates to what may be termed "metering" an event or "capturing" a sale. The second bullet more roughly relates to what is called "billing" although it also covers updating of a customer's balance, a description more appropriate for prepaid customers. Charging integrity means that the supply is correctly recorded, and the consequent charge is correctly rendered.

What Measurement Means

The verb "to measure" means to

- ■ Ascertain
- ■ The extent or quantity [of a thing]
- ■ Compare to a known standard or fixed unit

This also means that measurement excludes the following:

- ■ It is not guesswork or a subjective judgment.
- ■ It cannot be expressed in qualitative terms, as the aim is to represent in numbers the relationship between the thing measured and the unit of measurement.
- ■ It is not subject to reinterpretation, as the standard or unit does not change.

The Thing That Is Measured

Measuring integrity is the same as measuring its opposite, error. What complicates things is that the number of errors may be uninteresting, because we really want to understand how serious the errors are. Some errors may have no impact, but others may have enormous consequences, and it can be difficult to extrapolate from the kind of error to the likely impact. The purpose of this measurement approach is to quantify the impact of all errors that affect the final charges levied. These errors are determined by comparison to what should have been the charged.

It may be thought that it is necessary to know what the errors are before they can be quantified. That is one approach, but knowing an error has taken place need not mean the error itself has been identified. Imagine a situation where a customer in a shop receives the wrong change. The person serving them may have incorrectly keyed the sales into the till, or added up the change wrongly, or the till may have been in error, or the bar code on one item may have been associated with a price different to that stated on the product's label. Whatever the reason, the important thing is that, just from scrutinizing the change and without looking at the receipt, the customer can tell a mistake has occurred without knowing exactly why. There is hence a difference between

- Detecting and measuring error in general
- Investigating and analyzing the causes of error

This article is about the former, not the latter. The emphasis here is to quantify error correctly. As a result, the techniques described do not assume that certain errors are likely or unlikely. Any measurement technique that presupposes the outcome is biased. The techniques described are intended to quantify the extent of error correctly without prejudging the outcome.

Numbers and Units

A measure is ultimately always expressed as a number and a unit. Clarity is needed as to what are the relevant units for measuring the integrity of charging. The two types of units that are primarily used in measuring charging integrity are as follows:

1. Units of monetary value
2. Counting discrete events and charges

Given the purpose of measurement is to assure accuracy of charging, it follows that the extent of error in determining the value of charges presented is the key numerical property to be determined. Customers and businesses are likely to be most concerned about the monetary value of errors that take place more than they are the number of items that are in error. However, it is also worth considering how many discrete charges are in error. This is because the cost of correction or the impact on customer perception may be more closely related to the number of individual items in error than the value of each error. This is most obviously true when comparing a large number of errors, each of small value, to a small number of errors of large value.

Depending on how a charge is calculated, the business will need to record properties expressed in other kinds of units, such as

■ Time
■ Data volumes

Quantifying integrity may hence mean that the accuracy of recording properties in these units must also be measured. However, this is subordinate to determining the impact in terms of value or numbers of incorrectly rendered charges. Hence, the appropriate units for measuring error in terms of time, data volumes, and the like depends on the way charges are calculated and the rates applied.

Defining the Measure

The philosopher Wittgenstein commented that measures must be consistent with each other, and that otherwise the whole system of measurement fails (Wittgenstein 1983). However, many RA functions have unwittingly implemented "measures" that cannot be said to be consistent with each other. One of the goals of this article is to articulate how to avoid falling into the trap of generating incomparable, and hence questionable, measures. The first step in ensuring comparability of measures is with the definition of measures. Measures should be clearly defined. Definition also clarifies what is not being measured. By defining a measure, it is also possible to state what level of confidence can be obtained from any given measure as compared to any given target. To avoid the need to revise a measure, it is a valuable discipline to state a clear definition at the outset and stick to it. It is undesirable to have a situation where the measurement is in need of revision, as this undermines the original measure's credibility. Measurement is based on fixed standards, so measurement methods should not change on a regular basis. Changing the basis of measurement also impedes comparisons between results drawn before and after the change.

There is not much purpose to measurement if the results are subject to perpetual revision. If there is reason to doubt the validity of the measure, the method, not the output, needs to be changed. It is inappropriate to revise a single result in isolation, because this introduces a risk of bias. If results indicate a problem with the method of measurement, then the method should be corrected and the measure collected afresh and consistently from then on. Where a measure has been correctly obtained but is flawed by chance, say because a sample is not representative, it should still be necessary to demonstrate the flaw instead of just assuming it. This might require the collection of further and larger samples, or analysis of the cause of the error to show that the real likelihood of the error is significantly different to the results obtained from the sample.

Any basis of measurement that permits results to be disregarded when they do not fit preconceptions is flawed. Any basis of measurement that ignores errors on the basis that they can be treated as "resolved" because of subsequent actions is also flawed. Either of these so-called "corrections" of measures should be properly considered as forms of bias. The measurement approach stated in this document provides for no post-measurement revision of results, with the sole exception of methods to address sampling error.

A measure works well if it can be applied consistently in different places and at different times. It serves no purpose if each measure is itself a one-off. As the purpose is to establish the integrity of charging, there is no value to a measurement approach that permits each new measure to be determined on a one-off basis without consistency.

It is reasonable to expect a degree of trial and error when establishing measures of charging integrity. However, that period of trial and error should not last indefinitely. A flaw in the method of measurement will typically imply the results obtained using that method are also flawed and should be disregarded. Hence, the objective is to design and implement measures that are free from any flaw.

Appropriate Degrees of Precision

It is reasonably obvious that to measure to a certain level of precision, it is necessary to implement devices, processes, and the like capable of measuring to at least that level of precision. The precision of the measuring device depends on how it works. For equipment, like test event generators, attainable levels of precision should be documented in a reputable manufacturer's specifications. For example, the test event generator will measure the accuracy of an event's duration within a certain degree of accuracy. Sometimes it will be suitable to test the precision of the equipment independently.

Although discussions of precision usually focus on automation, it is also important to consider human elements that may affect the reliability of results. For example, a person required to manually inspect an invoice is unlikely to find all the errors if given insufficient time. This is not a reason to object to a human element to measurement, but human factors must be allowed for. There are advantages to using people rather than automation for measurement, especially where automated checking is too expensive or inflexible. Just as it is necessary to test and monitor automated checks to ensure they are correct, similar techniques can be applied to manual checks. Such techniques include varying the people who perform checks, performing cross-checks or reviews, and training people to be aware of changes before they take place.

There may be other factors affecting the precision of measurement unrelated to how the measurement device itself works. These factors may be described as environmental. They relate to a fundamental (though usually very small) inconsistency between what is being measured and how it is being measured. A simple real-life example would be to measure the time at which a signal occurs from the point of view of the source and the destination. However quickly the signal travels the distance between the two points, there is some delay. If the measurement device is located at the signal's origin, but is used to measure when the signal is received (or vice versa) then the delay may affect the level of precision in measurements taken with that device. Another example is the delay to the processing time of CPUs that results under increasing load. If the delay can be precisely determined, and is consistent, then it may be allowed for without a reduction in precision. If the delay is negligible, then it may be ignored. However, if the delay cannot be precisely determined, is variable, and cannot be considered negligible, then it will reduce the level of precision in measurement.

Just because a delay (or similar property) may be determined in theory, this does not mean that the relevant scenario in the real world has been shown to be consistent with the theory. For example, although it is possible to determine how long it takes a signal to travel between two specific points, it may be impractical to determine or even estimate the delay where signals may travel a variety of paths between a large number of combinations of source and destination. Although the delay can be precisely measured for any two points, it is more reliable to make allowances for the range between the shortest possible and longest possible route that may be used if it is impractical to relate the exact delay to each specific event. The wider the range, the greater the imprecision that results from this environmental factor.

When measuring performance by counting discrete numbers of events, the more events that are counted, the more precise the measurement. For example, suppose a sample of 10 events is observed, and 9 are found to be correctly charged. The error rate in this sample is 10%. Ignoring the risk that the sample is unrepresentative, the most reasonably inference is that the population error rate lies between 5% and 15%. Probability tells us that 9 out of a sample of 10 is the most likely outcome for a population error rate between 5% and 15%. To be more precise in estimating the population error rate requires a bigger sample. If the sample had been of a 100 items, with 90 correctly charged, the population error rate could be inferred to be between 9.5% and 10.5%. Ignoring the risk of unrepresentative samples, it is true that the more data there is, the more precise the ability to measure the rate of error. So, it is wrong to claim higher levels of precision than permitted from the volume of data obtained. For example, if you only test 10 items and find no errors, then you cannot deduce the risk of error is less than one in a hundred. The issue of precision is most pertinent where the expected level of error is very low. If an error is expected to occur only once in 100,000 events, and it is not possible to separately monitor the causes of error in that one case, then at least 100,000 events need to be inspected to conclude whether 1: 100,000 is the error rate. Allowing for sampling error further increases the amount of data needed to obtain a conclusion to the desired level of confidence.

When formulating a measurement method, it serves little purpose to attempt to measure to a level of precision not permitted by the following:

- The accuracy of measurement equipment and processes
- The quantity of data available
- Environmental factors

It may be difficult to assess these without trial and error, but once assessed the expectations of measurement must be set at a realistic level. If not, results will not be meaningful.

A good analogy to the aforementioned is that of a marksman shooting at a target. The closer the marksman is to the target, the smaller the bull's eye he may realistically intend to hit. If he stands further away, the same deviation in aim will mean a miss instead of a hit. Also, standing further away means he must pay more attention to environmental factors like crosswinds. The variability of these reduces the accuracy he can achieve. This means his realistically attainable target must be larger at a greater distance. When devising a measurement approach, it is worth considering how "close" the measure is to the property being measured, and set the measurement expectations accordingly.

Error Ratios

One consideration in the measurement of error rates is how that error rate itself is to be expressed. Avoiding confusion becomes especially important where higher levels of precision are expected. It is likely that error rates will be expressed as some form of ratio, which of course requires a numerator and a denominator. This error ratio will reflect the number of errors expected to be identified in a representative sample of a given size. For example, there may be 1 event in error per 1000 events, or $1 aggregate error per every $1000 charged.

The numerator in such a ratio is straightforward. This will be the extent of error identified. However, an element of choice exists around the selection of the denominator for an error ratio. The denominator might be either the total including errors, or the total excluding errors. In other words, it might be the total before identifying errors (as might have been originally recorded or presented), or the total that should have been presented if error free. As equations, the choice is between

- (Errors/total of population including errors)
- (Errors/total of population excluding errors)

If the calculation takes a denominator that excludes error, then we compare the errors with what should have been the correct value or number of transactions. In many ways this is preferable, as it eliminates the bias introduced by the error itself. For example, if customers are overcharged, then including the overcharge in the denominator makes the fraction smaller, and thus gives a more flattering impression of the error rate. However, where the intention is to calculate on the basis of the total excluding errors, it should be recognized that this figure can never be determined with complete certainty, because there is always a risk of errors that have yet to be identified as errors.

Consider this hypothetical example of a test for overmetering. 1000 events form the test sample, 1 of which has been duplicated (and thus overmetered). Therefore, 1001 events have been recorded, although only 1000 events actually took place. Suppose this is the only error that has been identified within the sample. The error rate can be expressed in one of two ways: either 1 error in 1000 events or 1 error in 1001 events. The difference in the error rates based on totals, including and excluding error is approximately equal to the square of either error rate. For example, the difference between 1/1000 and 1/1001 $\approx (1/1000)^2 \approx (1/1001)^2$. This means the importance of using a consistent denominator increases relative to the expected rate itself. As noted, the error rate as calculated using either denominator need only be squared to determine the total numerical impact of the decision to include or exclude errors in the denominator.

Breaking Down Charging into the Three "C"s

The terminology for describing how a charge is processed for presentation to a customer is confusing. Many processing steps may occur between the point where a supply is made and the point where a charge is presented. The sequence, number, and even types of step involved will vary between communication providers. Indeed, the accuracy of the charge may depend on processing stages split between two or more separate businesses, in situations where they pass data between each other. Terminology tends to be based on the names of the automated systems that comprise the chain of processing from supply to bill. Often there is no good terminology to describe the manual activities and workarounds that are also vital to processing. The term "billing" focuses on processes at the end of a chain. "Metering" might be used to refer to the start of that chain, where pertinent data records of an event are generated. There may be many discrete processing stages between these two points. Some of the words we might use to describe the automated and human activities that take place during this chain are recording, logging, mediation, provisioning, rating, real-time account queries and updates, guiding, and suspense recycling.

I treat charging as a holistic activity and define it accordingly:

> "Charging" means all cumulative activities that enable the sale of a supply to be recorded and then rendered as a monetary amount to be paid by the customer.

Charging integrity hence concerns an entire end-to-end chain of processing from recording what has been supplied to presenting the charge to the customer's account. Now we need a model that breaks this down into its logical components. Note that the logical components need not readily correspond to different discrete systems or processes, because it may be that several logical activities take place in parallel. In turn, a single logical activity may involve the operation of several discrete systems or processes acting in sequence.

Charging as Three "C"s

Processing data for charging involves the following three logical activities:

1. Capture: the recording of the supply that is made
2. Calculation: determining the charge to the customer for the supplies made
3. Conveyance: the transmission of data from the point of recording the supply to the point of presentation to the customer's account through all intermediary systems

This article discusses a model for integrity measurement based on the logical activities of capture, conveyance, and calculation. This will be referred to as the "Three 'C's model" or "CCC model."

CCC Model: Capture

The term "capture" is used here to cover activities commonly known as metering, logging, recording or the taking of an order. There is an analogy with an old-fashioned meter recording the amount sold. The analogy is helpful, up to a point. The basic purpose of capture is to generate sufficient data when a sale is made to correctly charge for the sale. This charge may be determined simultaneously, as with real-time prepaid services, or it may be calculated later, as in a classic postpaid billing architecture. Here are some simple examples of capture:

■ A voice call is made, and the switch produces a CDR that contains relevant parameters such as the origin and destination, the time the call took place and the duration.
■ A customer telephones a helpdesk to request a new service which incurs a one-off activation charge, and this is typed into the sales order system by the helpdesk assistant.
■ A customer orders pay-per-view television content by using the interactive menus managed by their set-top box.

Capture is different to calculation and conveyance in its nature because it is always time-critical. The data must be recorded when the relevant events take place. It is not possible to record a sale retrospectively if no data was produced at the time of the sale. Unlike conveyance or calculation, there is no possibility of reworking if there are indications of error. Even with real-time charging, it will still be possible to rework retrospectively for any errors in conveyance or calculation, so long as the original capture data is retained and is reliable. If there are reasons to believe that capture data is incomplete or invalid, it will be impossible to correct the data systematically afterward. The only way to do that would be, in effect, to implement a second, separate capture process to serve as a potential substitute. However, if the two separately captured records disagree, that only begs the question of which one is right.

One source of confusion is that the data output from the capture process may not be retained at the system used for capture for any significant length of time. In many real-life situations the data is immediately or soon afterward conveyed so it can be stored in a data repository elsewhere. This may complicate the investigation of capture errors if testing is based on the repository's data instead of the raw meter output; errors identified in the repository's data may actually be the result of failures in the process to convey and write data to the repository. As it is possible to construct a measure based on the direct capture output, it remains true to regard capturing as a separate logical activity from conveyance to the data repository. As with many aspects of measurement, we can understand a distinction between a theoretical goal of measurement and real-life compromises to get a reasonable approximation. It is important to distinguish between the theoretical ideal and practical compromises to recognize the implications of compromises.

Because capture must be real-time, and cannot be subsequently reviewed, it means that any measurement of capture error must be based on collecting real-time data from a comparable system. To check the capture, it is necessary to use an independent system that is provided with identical inputs and is designed to emulate its outputs. In effect, an alternative meter is needed to provide the test data. Any disagreement between the meter and test system begs the question of which (if either) is deemed to be in error, though the practical intention must be to generate a test approach where any discrepancy can be safely ascribed to error by the meter. As Wittgenstein also noted, we must keep an open mind about whether, when comparing results, we should sometimes also reflect on the integrity of the measurement system.

CCC Model: Conveyance

The term "conveyance" is used here because it is meant to represent the fact that the data output from the meter must be conveyed through a number of intermediary systems before being conveyed to the customer's account, whether in the form of a bill or other update to their account. There is also an analogy between conveyance and a manufacturing conveyor belt. In this charging production line, the data that records a sale is placed on the conveyor belt at one end, a series of processes take place in sequence to reformat, normalize, and add to that data. This ultimately leads to the production of data that is presented on a bill or similar, and which also states a monetary charge. The idea of conveyance may appear trivial and obviously error free. Because it takes place between systems and its complexity is often underestimated, it can be a poorly controlled source of very significant errors that go unidentified for long periods. Here are some examples of conveyance:

- CDRs are polled from an exchange and transmitted to a mediator.
- CDRs pass "through" a mediator, being input to it, normalized, and then output to the destination system.

- The billing system generates a print file, which is sent electronically to another business to print the bills.
- A provisioning system flags the completion of a previously outstanding order, and this flag is identified by a batch process causing it to update a record in the billing system.
- CDRs fall into suspense at the billing system, and then are successfully recycled.
- A salesman writes down an order on a paper form, which is later returned to another department to key in the details into the sales order system.
- Printed bills are sent through the post to the customer's address.

Some of these examples suggest an important question: what are the "ends" to this conveyor belt? Is it from the customer (at the point of sale) back to the customer (at the presentation of the charge), or is it enough just to consider the inputs and outputs of a single business, or even a single system? This question becomes even more pertinent when the activities of more than one business contribute to the "conveyor belt" for a given customer. For example, a switch at operator A generates a record of a call, and the operator passes the CDR through a regular electronic file transfer to service provider B to calculate the billable value, which then creates a print file and sends it to billing bureau C so it can print and send out the bill. Errors may occur at A, B, or C, but all will affect the charge the customer receives. A suggested practical definition of the end points is given below

- The output of the capture system (in this case, the switch at operator A) is the starting point for conveyance.
- The end of conveyance is the point where data is output in a form where any further processing is for presentation purposes only.

By definition, the capture system produces data intended for use in determining the charges to be rendered. This definition is somewhat arbitrary, but the point is to focus attention on the actual system or systems used to make a record of a sale where that record is the basis for subsequent charging. Other systems may also make records that could have been used as the basis of charging, or may even be reconciled to the record that is used, but for the purpose of integrity they are irrelevant. What matters is the correctness of the output from the device that does produce records used as the basis for charging. This is hence the most sensible starting point for measuring correctness of data conveyance.

A reasonable end point would be where any other processing to be performed should have no impact on the final value of the charges to be presented to the customer. Although it is conceivable that an error may take place during formatting that affects the charges presented, this would normally be a negligible risk. Of course, other conveyance errors may still take place, such as the bill being lost in the post. These kinds of "errors" are of a very different nature to the issues this kind of measurement is intended to address, but there is nothing to prevent measurement of these kinds of errors as well.

Note that some of the examples stated earlier would be excluded from measurement if the suggested end points are used. For example, losing a bill in the post falls outside the suggested end points. Whichever end points are selected, it is most pertinent that they are clearly identified and consistently used. Indeed, when explaining measures, clarity in the end points is necessary for the measures to be meaningful and properly understood.

Conveyance is not something that necessarily occurs before or after calculation. It is best considered as something that occurs in parallel with calculation. In the aforementioned example using suspense, a calculation could not be performed; hence, the CDR could not be conveyed to the next step. Instead, it is redirected to a holding data repository until it can be resubmitted (or "recycled" to use the normal terminology). This can be thought of as a small loop that comes off our conveyor belt and returns items back to an earlier point. The point here is that performing the calculation, so that we get the calculated output from the precalculation input, is actually part of the conveyor belt.

CCC Model: Calculation

The term "calculation" is not used here to mean any kind of manipulation of data. Data is often manipulated in many ways that do not in any way change the nature of the information relayed by that data. A change in the format of data is not a change in the information it conveys. Changing a number represented in, say, binary to the same number represented in base ten could be said to involve calculations of a certain type, but that is not how the term "calculation" is used here. With respect to charging integrity, calculation includes any operation necessary to determine a monetary value from the conveyed meter output. These operations may be performed at once or in a series of discrete parts. However, their cumulative output is the final monetary value of the charge due. Some examples of calculation are as follows:

- A CDR for a voice call is rated according to the event type, the A and B party, the tariff plan associated with the A party, the time the event took place, and its duration.
- All calls that a customer makes in a certain month are aggregated by the billing system and a discount applied that varies with the total duration of calls made in that month.
- A leased line is implemented, causing an update in the billing system, and from that point the line is charged at a standard recurring rate looked up in a reference table.
- Outages of a leased line are recorded in a database to determine if service levels have been met, and if service levels have not been met then a flag is

set causing the billing system to calculate credits to net against the recurring charge.

■ A customer phones a helpdesk and requests a new service, so the helpdesk assistant flags a new order in the billing system, causing a one-off flat fee to be looked up and applied to that customer's account.

■ A customer sends an SMS text message to a number that is recognized as charged at a premium rate, as it is a request for an automated SMS response of weather information.

■ A customer sends an SMS, which causes an automated SMS response of weather information, the receipt of which gets recorded as a chargeable event at a premium rate.

■ A certain number of events are allocated against an allowance each month, the consequence being that these are provided at zero charge.

Just as conveyance occurs between many systems, more than one system may be involved in calculating the value of a charge. It may even be the case that more than one business is involved in calculating the charge. For instance, operator A records a call and applies a charge, then passes it to operator B. Operator B has the billing relationship with the end customer. Operator B adds a margin to the call charge calculated by operator A, and presents the total charge, including their margin, to the customer. In this case, both operators perform part of the total calculation of the charge presented to the customer, and either or both might commit a calculation error.

It is not always the case that there is a simple relationship between each event and the charge incurred. This is most evident where there is a discount applied to the total value of a bill. There may be no means of calculating how much each individual event has been effectively discounted. We might think of the discount being netted against the standing charge, instead of against the charges for the individual events listed on the itemization.

Calculation implies augmenting the data captured when the original event takes place. Capture produces the necessary raw data that records the relevant particulars of a sale. Calculation produces further data that represent the monetary charges (or at least produces intermediate data that will be further processed by subsequent calculation steps). Calculation is use of an algorithm based on data relating to the captured events and the appropriate rate of charge for the customer. In practice, this means that customer data needs to be looked up to perform the calculation. The way data typically is organized, the customer will be associated with a specific tariff plan. For a calculation to be correct, both the algorithm and the standing data must be correct. These are correct if they are consistent with the basis for charging agreed with the customer. Any measurement of the correctness of calculation must at some stage link back to the price plan that has been communicated between the communication provider and its customer.

In one of the examples given earlier, the concept of a zero charge is described. This must be distinguished from the case where no event is presented on the bill because that kind of event is not chargeable. A zero charge implies that a charge has been calculated. Where no charge will ever be levied for a certain kind of event, it may be easier to avoid metering the event, or else decide not to convey the record of the event, instead of calculating a zero value for it. The case where a zero charge is calculated and conveyed to the bill needs to be clearly distinguished from the case where the event is simply not metered or conveyed to the bill. This is because there may be integrity reasons to present zero charge events on the bill. Though the event itself may not be charged, its inclusion on the customer's account may still alter the total charge to the customer. The most obvious example would be where a certain volume of events is subject to an allowance. A call covered by the allowance incurs no charge, but it reduces the remaining allowance for subsequent calls, making it more likely that later calls will incur a nonzero charge. Not conveying a record of an event to the bill is only acceptable if there could never be an affect on the end value of the bill and the customer can be safely assumed to have no need for this information.

Concept Diagram for the End-to-End Charging System

Figure 5.5 represents the concepts described earlier.

There are other diagrams that try to explain how charging works from end to end. However, these usually illustrate specific systems and how they relate, in ways that may or may not be true for any given communication provider. For some services and some communication providers, a CDR is generated by the switch, polled and then aggregated and reformatted by mediation, rated by the rater, and then sent to the billing system, which applies discounts, standing charges, and then presents the total owed for the month. For many other services and communication providers, the charging chain will be very different. For example, a prepaid charging sequence will typically be very different to that found for postpaid services. As a consequence, supposedly generic pseudosystem representations of charging chains are often unhelpful. They do not reliably express the concepts that are key to measurement. Very abstract system diagrams still give a very simplified view of how real-life systems interact, and are unlikely to give an adequate and consistent basis for forming a measurement model. Real-life system architectures will always deviate from any attempt to describe typical systems operations. Instead, it

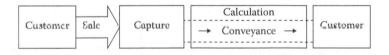

Figure 5.5   The end-to-end charging system.

is preferable to start with a clear view of the principles, and then map these to the real-life systems. That way, the concepts always give a solid and consistent foundation for measurement.

Error Implications

Table 5.2 gives a simple practical analysis of the different implications of an error, depending on which of the Three "C"s it relates to.

Black Box Measurement

A black box view is so-called because no knowledge is assumed about how systems and processes actually work. All that is known is the initial input (what was supplied) and the final output (the charge to the customer). Hence, all we can do to monitor is to compare what came out with what went in. The comparison is based on our expectations of how the two should be related; in other words, what the pricing schema is. Black box measurement works solely based on our expectations of how things are supposed to work, without our needing to know anything about the detail of how things actually work in practice. All we can do is base our measurement on whether the data output (the charges and any itemization presented to the customer) is correct given the input (what was supplied to the customer).

White Box Measurement

A white box view involves having knowledge of not just the initial inputs and final outputs but also the details of how the systems and processes work between that initial input and final output. The box is white, as opposed to black, because we are

Table 5.2 CCC Model Error Impacts

	Capture Error	Conveyance Error	Calculation Error
If an error is committed, but sufficient capture data is retained, will it be possible to reprocess the capture data and retrospectively correct the errors?	No	Yes	Yes
Would a customer be able to identify the error by checking their itemized bills against the relevant tariff?	No	No	Yes

able to see inside it. Because we can see inside, we can devise an approach based on the details of how things work in practice, as well as relating to the end-to-end expectations.

Using a white box approach, the definition of measurement can come from, both, how things should work, and how they do work. Instead of just measuring between the end points, it is also possible to measure between intermediate points and hence to cumulative add measures. Measures can be devised to place more emphasis on determining the scale of the likeliest errors, where likelihood is determined by assessing how things actually work or is based on past experience. For example, if the charging chain involves just three discrete steps, we would add up the errors we detect in steps 1, 2, and 3. Furthermore, we may do only very low levels of testing for steps 1 and 3 because preventative controls and the simplicity of processing make errors very unlikely. In contrast, step 2 may be known to be very error prone, so we may do a lot more testing of this step. In that case, we do not simply add together the test results but weight them according to the size of samples. For example, if step 2 is subject to ten times as many tests as step 1, we would scale up the number of errors found in the sample taken in step 1 to give a fair weighting relative to the end-to-end charging chain.

In a modern communications business, there is often very significant complexity in how systems and processes work. This complexity may also be exacerbated because the role played by a system as part of the charging of one kind of service may be quite different to the role that same system plays when involved in the processing of charges for another kind of service. It may take a significant level of resource just to identify the relevant measurement points for white box testing, and even more to implement measures across all of them.

Measurement in a Black Box or White Box Context

In Figure 5.6, we see that a black box measure is based on just the main data streams that go into and come out of the black box. In a white box measure, we can measure performance between each of the points within the various systems and processes, and then combine these to get a cumulative score.

Both black box and white box measures should give the same results. However, there are numerous practical reasons why they may give different results, which can be summarized as follows:

- An incomplete white box approach may miss some of the steps necessary to get a truly complete measure.
- A black box approach may be overly simple and hence exclude certain real kinds of error from measurement results.
- A white box approach may not correctly weigh each of the results before it is combined, exaggerating the importance of some results but undervaluing the importance of others.

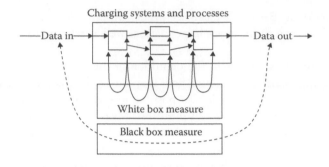

Figure 5.6 Black box and white box measurement.

Let us compare the relative strengths of the black box and white box approach to integrity measurement, in detail, beginning with the black box approach. A black box approach to measurement is as follows:

■ Easy and quick to understand and audit because it requires no detailed knowledge of real-life systems or of relative risks
■ Easier to maintain because it can remain valid even when the measured real-life processes and systems change
■ Less likely to omit relevant data than a white box approach
■ Better at providing high-level metrics without the need to appropriately weigh and combine data from many sources
■ Easier to reproduce in different contexts because it is not related to the underlying systems and processes
■ A useful high-level "barometer" especially if the results are produced on a timely basis
■ Not capable of giving a misleading partial measure because of gaps in the data
■ Not subject to some of the errors that may occur with a white box method because it does not require a comprehensive understanding of systems and processes
■ Easier to adjust for changes in tariff policy, service offerings, and customer behavior
■ Intuitively easier for a customer or auditor to understand, because the measurement is based on what should be charged given what was supplied, irrespective of how this is done

In contrast, a white box approach

■ Gives useful information even when incomplete
■ Can be assembled by combining existing sources of data and filling gaps in existing data rather than implementing a completely new measurement approach from scratch

- Enables most effort to be focused on the likeliest sources of errors instead of just providing a lot of generally uninteresting and repetitive results
- Puts less onus on devising a comprehensive understanding of tariffs, customer behavior and other factors that alter the relative importance of errors across the end-to-end charging chain
- Is less subject to macroscopic errors due to inappropriate or unrepresentative samples
- Should encourage a sufficiently complex view of how systems and processes actually work that there is less risk of errors due to oversimplification
- Is more useful in diagnosing the cause of problems because it works at a lower level
- Forces a thorough understanding of how systems work and are integrated
- Forces a thorough understanding of where the greatest risks lie
- Will provide data of more immediate use and relevance to specific parts of the business

In general, a black box approach is very well suited to high-level measurement across a whole business or across a chain of related platforms. The simpler the tariff plans, customer options, customer behavior, and the like, the easier it will be to use a black box approach to assess the relative impact of error in a simple and intuitive fashion. Black box approaches become less useful when the nature of charging is so complicated that it is no longer possible to devise simple end-to-end measures. For example, the more complicated the variety of sales recorded, and means of recording them, or the more complicated the charges and means of rendering them to bills of different formats, the harder it will be to perform checks that cover a broad range of outputs.

A white box approach is better suited to a business that already has many of the relevant sources of data but needs to tie them together. One major disadvantage of the white box approach will be that it is much more costly and difficult to implement. The cost may, however, be reduced if the measure can build upon existing piecemeal detective controls. But unless all the processing steps are covered by the white box measurement activities, so there are no gaps in measurement, the white box approach will degenerate into nothing more than an incomplete inventory of issues. Like any incomplete inventory, the total result will be understated, and no conclusion can be formed about the significance of items missing from the inventory. A real advantage, however, is that the data delivered by white box measures is much more relevant to a given system or process, meaning it can still be productively used in the absence of other measures. They are also much better suited to diagnosing the cause of problems because each individual submeasure has a much lower span than the wide-span black box measure.

In reality, white box and black box are two ends of a spectrum. White box measures are low level. They give better data at that level but with white box measures it is harder and costlier to generate a comprehensive view of performance. Black box measures are high level, which increases the risk of oversimplification, but also gives a more intuitive overall understanding of the relative issues in performance. The one caveat for black box approaches is that it will often be necessary to compromise to get working measurement technology and human processes because of the difficulty of putting a single genuinely end-to-end measure in place that will test a high-enough volume to give reliable measurement results.

Combining Black Box and White Box Measurement

For communication providers that want both reliable measurement, an effective response to any leakages identified, and feedback on the quality and power of the controls they have in place, the best approach is to implement complementary black box and white box end-to-end measures of integrity.

- The black box approach is high level and is designed to produce a single end-to-end measure on an ongoing basis. As such, it is the definitive source for overall compliance measurement and for assessing the total of leakage.
- The white box approach is low level and is designed to record each and every detailed issue identified from low-level monitoring, investigation, problem management, and other such details. It is a reflection of a primarily diagnostic approach to integrity. Each submeasure can be separately revised and updated. This revision will be necessary whenever there is a change to the system or process spanned by a submeasure. The submeasures create an inventory that, if added together, can be summated to an end-to-end measure. However, there is no guarantee that this inventory will ever be complete, even if it seems complete. This is because there may be gaps in the inventory that are not appreciated. This risk of gaps is compounded by the fact that there are diminishing returns in addressing what are judged to be increasingly improbable risks of error. As such, white box measurement is an inferior method for generating reliable and recurring compliance measures to be reviewed by regulators or external auditors (who might reasonably question subjective judgments of risk in the absence of empirical data either way).

The two measures can be independently generated and then cross-checked for greater confidence. Although the white box measure is not designed to be as precise or reliable a source for end-to-end measurement as the black box measure, cross-checking has the following virtues:

- Any oversimplifications in the black box model can be identified from the detailed line item analysis in the white box model.
- Any gaps in the white box model may be evidenced by differences with the end-to-end measure generated by the black box model.

- One or other model may be better suited to rapidly cope with any given system or process change, so even where changes take place the time for which there is inadequate measurement is minimized.
- Major discrepancies may indicate flaws in either source of data or the means of calculating their impact in measurement terms, so comparison is a broad-span control over quality of data and appropriateness of measurement calculation.

The obvious disadvantage of using both measurement approaches is that there is increased cost to the communication provider. This may seem wasteful if the business just considers the independent measures as providing essentially the same information. However, this understates the advantages of supporting tight diagnostic controls with a checksum that highlights any gaps between them. The black box approach is based on a very high-level view of the business with the fewest possible points of data analysis to minimize the effort required. The white box approach, on the other hand, involves a lot more detailed work to compile, though the actual figures are a secondary output from much more detailed monitoring and investigative work. For a well-controlled communication provider with active response times to issues identified, that tightness of monitoring is necessary anyway.

In practice, black box measures will be the output of a series of equations with variables populated from raw data obtained from high-level capture, conveyance, and calculation checks. The equations enable the probabilities and significance of each kind of error to be appropriately weighted to give a representative total error. In contrast, white box measures are usually given in the form of relatively extensive reports (though these may be summarized in practice) of all the various, separate checks that are performed. These measures more specifically highlight critical and major issues and their impact on the business. The appropriately weighted white box submeasures are totaled so they give an end-to-end measure that can be compared to the black box measure.

Real-Life Measurement of Capture

To test and, hence, measure the systems and processes used to capture, it is necessary to identify the source that generates raw records used for charging each sale. This may be far from straightforward, and naïve RA practitioners often underestimate the significance of this challenge—sometimes to the extent of monitoring the wrong systems. In the following discussion we use the term "meter" to refer to any system or process that generates the original capture data of a sale.

Only One Meter for Each Event

Not everything that may capture data of the sale is the meter. This is because more than one device may produce equivalent data that could be used as the basis of charging. The important point is that, where there is a choice of potential metering systems, it still remains true that only one is actually intended to be used as the source of data for charging. Data from other systems, not intended to be used as the basis of charging, are irrelevant. However, such data may be used as a control over the quality of the meter, for example, by reconciling the two outputs. For each event reviewed, it is necessary to identify which device produced the relevant data and check that output. The correctness or otherwise of data output from other devices is ultimately irrelevant to charging integrity.

A simple example would be a case where a customer leaves a voicemail on an automated platform. The switch that manages the call, which is automatically redirected to voicemail, may produce a record of the call just like any other call. The voicemail platform may also produce a record of the delivery of the message. Either the record from the switch or the record from the voicemail platform may be used as the basis of charging. In real life, the communication provider may use one but not the other. It is possible to imagine a case where the customer receives a charge for making a call (like any other) and for depositing a voicemail where the former is charged per the switch record and the latter per the voicemail record. This may even be presented on the bill as one charge for one event. In this case, it is best to treat them as two separate events, each with an independent and different meter. This still enables meter errors to be distinguished from conveyance and calculation errors.

Only One Meter but Maybe More Than One Device

A system-based view may lead to the impression that one meter = one device. This is not precisely true. In the conceptual model, the meter produces the data. Though we talk about "the" meter it may be that more than one device produces data where all the cumulative data is needed to calculate the charge. This is not the same as the aforementioned example, because in this scenario each device is needed for metering because each produces different but necessary data.

A simple example would be where a call takes place and the duration and start time are recorded by two separate devices. Call data is passed via an intelligent network device to a real-time billing solution. The start time for the call is recorded by the intelligent network device at the start of the call. That datum is conveyed to and stored by the real-time account management solution. The duration for the call is recorded by the switch at the end of the call and passed via the intelligent network to the account management solution. In this case, the device that records the start time and the device that records the duration are different, though both have their data output conveyed to a third device, which uses all the data to calculate the final charge. One source of confusion here is that the metered start time is produced by the intelligent network, which also conveys the metered duration. On the other hand, the switch may have been able to record both start time and duration, but in this instance the start time recorded by the switch is not that used for charging the customer. An attention to detail is needed to identify correctly which device produces which datum as part of metering. However, it remains true that for each datum needed to charge an event, only one device produces the datum actually used as the basis of charging.

Contingency Meters

It is possible that, where a choice of meters exists, one may be used as a substitute should the usual meter fail, for whatever reason. This is the same as saying that another device becomes the meter on a temporary basis. Any reliable RA measurement should reflect even a temporary change of meters. If test devices have been deployed, it may be tempting to use these as contingency meters as they may also produce equivalent data for the same events. However, where they are being used as the meter, it is no longer true that they can also produce a measure of meter correctness. The measure of meter correctness requires use of a second device that must be completely independent of the meter. If a test device is temporarily used as the meter then it cannot also be considered to produce measurement data about its own correctness.

Timing of Metering

Metering must take place at the time when a sale is made. However, this somewhat simplifies what might occur in real life, where the decision as to what exactly was sold might not occur just at one point of time. It might be that a sale is recorded, and then the details of the sale are altered or updated to take into account subsequent events. In this case, clearly all data needs to be recorded at the relevant point in time. This also means that output of several devices may be needed.

One simple example relates to service levels. Suppose a fixed line is leased on the basis of a guaranteed service level. Failure to meet service levels implies credits will be given to the customer. A sale is recorded when the line is procured. However, performance against the service levels is recorded on an ongoing basis. In other words, outages are recorded as they occur and then at some relevant point in time compared to service level commitments for determining whether a credit is due. So metering of what was in practice sold may occur at more than one point of time, because the original record of sale must be supplemented by data on subsequent service levels achieved.

A more complicated example involves a situation where a sale is recorded where it is known that the sale may not be completed satisfactorily and hence will need to be completely reversed. Although no sale takes place as such, two accurate pieces of data need to be produced at different points in time to record this nonsale correctly. For example, consider a situation where a text message is recorded when the customer sends it. However, the customer is not charged per every attempt to send a message but per every successful receipt of a message by the B party. Some time later, a message is received by the sending network indicating that the message was not successfully received by the B party. Metering has taken place at two points of time, by two separate devices. Furthermore, the devices may not have been operated by the same business. However, their correct cumulative operation is necessary for accurate metering.

The following question may be raised here: is the customer being charged for one thing in these cases, or is it that there is a charge and a netting refund, or two things that balance each other? The question may be thought academic, as the net effect, in terms of value, is the same either way. All devices that meter, at all points of time, need to work correctly to get the correct record of events for charging the customer. However, there are two reasons why this question might need a definitive answer:

1. The customer's perception of how many events took place may not be the same as that of the communication provider. In particular, even if netting figures have been correctly presented to the customer, it still is likely to lead to confusion if the customer does not expect to see netting figures. This is exacerbated if the netting charges are not presented at the same time to the customer, or if they are presented in such a way that it is not obvious they relate to each other.
2. Where a business or compliance target is stated in terms of number of correctly charged events, it clearly matters whether cases like these are counted as one or two events. It matters not because the question is interesting in itself, but because whatever rule is applied it must be applied consistently to get a consistent basis for comparison with the target. This is particularly relevant when dealing with compliance expectations of regulators or contractual commitments made to commercial partners. A flippant view that sets targets per event, but fails to define properly what one event is, risks an unnecessary dispute at a later date.

To illustrate the choices, here is one possible set of rules for how to count the number of events in complicated scenarios:

- Where the intention is to present a single figure on a bill, then one event has taken place.
- Where a charge and a credit relating to the same service is presented separately, the charge based on certain meter output and the credit based on other meter output, then two events have taken place. An example would be a charge for a message and refund because delivery was unsuccessful as presumably these are based on different, unconnected data.
- Where a charge and a credit relating to the same service is presented separately, the charge and the credit based on the same meter output, then one event has taken place. An example would be a charge for an event and a predetermined discount for that same event, as this is just a presentational nicety. Both elements are calculated from the same raw data.

A Practical Approach to Finding the Source of Capture

Capture takes place at the start of the processing chain. The paradoxical consequence of this is that to determine the meter, we need to first have an understanding of the other "end" of the processing chain, where there is the final output in terms of the monetary value to be paid by the customer. What needs to be recorded at the start is defined by how the charges are calculated. Hence, what needs to be captured cannot be determined in isolation from the subsequent processing stages. The best technique for identifying the correct source of capture is to trace back chargeable events from the final presentation through the various calculations and intermediate systems back to the original sources for the raw meter data. Only by tracing backward from end to start is it possible to prove that all relevant meter outputs have been identified. By tracing backward, and hence understanding the subsequent processing of the capture data, it is also possible to get a complete picture of what data needs to be captured for charging purposes (and what data is not relevant). Another benefit of the trace back technique is that it guarantees that no RA effort is expended on irrelevant data outputs that seem like meters but are not, in fact, the meter.

Identifying Conveyance

The trace back technique described earlier for identifying the capture source is also an effective way to determine the route by which records are conveyed. The key points for identifying the path for conveyance are as follows:

- There are many potential "end points" for a conveyance process, so to be sure of covering the right one, it is necessary to start with the end point of interest, for example, the bill or the customer's account, and then to trace backward methodically to the source of capture.
- Some processes involve the deliberate creation of duplicates or near-duplicates of records so they can be sent to different downstream systems (i.e., to meet different objectives). Effort needs to be methodically focused only on those versions of a record that are being covered by the intended measure. Whereas it may be practical from a diagnostic or controls point of view to cover all outputs from a system at the same time, this is not helpful for end-to-end measuring if it means effort is wasted on outputs that will not be measured any further.
- Conveyance is the transmission of relevant information, not of any and all data in any particular format. This means that attention must be paid to which transformations of data involve a change in the information (i.e., calculation) and those that just give the same information in another way (such as normalization). New information added, where relevant to the final value, must be correctly conveyed from the point it is generated. However, some

information may be added for reasons that are irrelevant to the final value of the charge. The correctness or corruption of this information is not part of measuring charging integrity.

■ Conveyance may be thought of at a "one record in, one record out" type level. This analogy is helpful but not always reliable. Several partial records may be combined to make one new record, or one record may be split into several, all of which may need to reach their correct destination for the final charge to be correct.

■ Though measurement of conveyance should be thought of at the level of individual records, data is normally conveyed in files that batch together large numbers of events. Many controls will hence focus on whether a file gets from one point to the next. However, this may obscure the underlying issues of conveyance. A file may be successfully transmitted without each record being successfully transmitted. Records may be omitted or corrupted, phantom records added to the file on its creation, the file may itself be corrupted or the file may not be properly read by the next system that receives it. Ultimately any measurement of conveyance must work at the level of individual records and not of files. Otherwise, certain kinds of conveyance errors will go unmeasured.

■ Loss of data is a typical conveyance error. Sometimes loss is real, whereas sometimes the data is not lost, but caught in a suspense or error log. Data is lost if it simply is not recorded anywhere, as may happen if memory buffers are exceeded, or where records fail a process but are not written to any suspense file or error log. No error has yet taken place if processing is merely delayed and the records can be resubmitted and processed successfully. Per the conveyor belt analogy, the items came off the belt and fell into a bucket to be returned back to an earlier point on the same belt. If, however, the items are not resubmitted, then they are effectively lost and a conveyance error has taken place. This is still an error even if the result of a conscious decision, because if the records represented real sales, then the business has failed to charge for all its sales correctly. In effect, the commercial decision is to commit an error because that is more cost effective than being accurate and because the customer's interests are thought to be unharmed.

■ Some loss of data may be deliberate, and not all loss is an error. A device may output records that are known to fall into error or are where there is no intention to pass them to any downstream process. The important point is whether these should have been passed on, which is why working back from the end is a helpful and practical way of viewing conveyance. Another form of loss would involve getting rid of irrelevant fields in a record. Again, there is no integrity issue if the lost field would not have affected the final value. The last form of loss is loss of precision, most notably in the form of rounding. Rounding is a perfectly acceptable form of loss of precision in data, as long as it is correctly understood. From a consumer protection standpoint, the customers should understand all of the ways by which rounding might affect the final value of what is presented to them. From the standpoint of measurement, rounding needs to be allowed for as part of the conveyance model. For example, if a system is expected to round a certain way, and does not do so in a given situation, then the failure to round correctly is itself a form of data corruption.

Identifying Calculation

As with conveyance and capture, calculation is best identified by tracing back from the end to the start of the processing chain. Calculation should be easier to pinpoint because it is unlikely to occur at more than one or two stages in the processing chain. Because calculation usually involves looking up standing data, it should be fairly apparent where this occurs. It should also be evident where new data is being written to represent a monetary value. The complexity in measuring calculation relates to ascertaining that the standing data, as well as the algorithms are correct. See below for more details.

Reperforming Calculation

To reperform a calculation, it is necessary to

- Emulate all the algorithms used in calculation
- Base calculations on equivalent customer data
- Base calculations on equivalent tariff data

All three pose different problems. To begin with, the precise emulation of algorithms necessitates a high degree of care. Care must be taken that calculations are properly understood and are correctly replicated. Otherwise, apparent errors identified by reperformance will, in fact, represent limitations of the algorithms used in reperformance. A seemingly mundane question that follows from this is the decision as to what, exactly, is being replicated. There is, in fact, a choice:

- The algorithms as designed and implemented in real life
- The calculation as described to customers

These should of course be the same. In practice, differences can occur. Choosing to emulate the calculation as presented to the customer means that the measure also covers any discrepancies between what is communicated and implemented. The additional difficulty with this approach is that any vagueness, inconsistency, or ambiguity in what is presented to customers, or any subsequent misreading by those responsible for implementing the check, may lead to the implementation of an unreliable algorithm for checking. However, the identification and resolution of issues of this type is a useful exercise in its own right. If the alternative approach is taken, and the check is based on the algorithms already implemented, the algorithms used for measurement should not be a direct copy of the code used in real life, as the results will always be identical. All that would then be tested is the data, not the algorithms themselves. It is still possible to cover consistency between the wording presented to customers and the algorithms implemented without basing the comparative algorithms directly on the wording. However, no measurement as such will be performed of the difference between what is stated to customers and implemented in real life. Instead, a control will be needed to ensure the prevention of discrepancies from occurring, rather than to measure discrepancies as they occur. This is a sensible deviation from a purely measurement-based view of integrity.

If consistency can be enforced using a preventative control more efficiently than errors can be detected, it is obvious that efforts should be focused on prevention. For example, having a single formal procedure for the pricing function to communicate tariff requirements to the departments responsible for their implementation and testing serves as a useful way to eliminate some of the most common types of error. These tariff changes would ideally be recorded in a single database to facilitate auditing. Consistency is enhanced by the departments responsible for implementation using an unambiguous specification that accurately represents Pricing's view of the requirements; and the Pricing department working with the assistance of the Legal department to ensure that the wording presented to customers is an unambiguous and consistent statement of what was formally stated in the requested tariff change.

The chief drawback of measuring by reperformance is the effort needed to implement a second mechanism to reproduce calculations. Given that the activity is intended to find flaws in the usual algorithms, it will always be questionable whether it is efficient to try to produce separate and correct algorithms to compare them with, rather than to apply additional resources to checking and improving the detail of the original algorithm. Though the algorithms need to be reproduced to enable measurement, it should not be assumed that they can only be reproduced using automation. A manual check of calculations is also a kind of reproduction. The best approach will vary depending on the relative cost and difficulty of implementing an automated or manual check. Sometimes manual checks can be very cost effective; people are very adaptable and may more easily emulate complicated algorithms and pricing structures. The best approach will also depend on how flexible or varied the reperformance needs to be. This relates to how much variation there is in real-life calculations and how often the tariffs change. People can change more rapidly than some systems. The cost of implementing and updating an automated system needs to be compared to the cost of employing and training people. Though there will always be some degree of human error, errors will also occur with an automated approach if insufficient resources are applied to the implementation or testing of the automated test device.

A sensible compromise is to reproduce calculations on an automated level where it is easy to do so, and rely upon the flexibility of manual checking for other aspects of calculations. For example, all calculations may be reperformed using an automated, but limited, tool. Where an exception is reported, this may be manually inspected to determine whether the difference reflects an error in the real-life calculation or a limitation in the reperformance tool. A slightly different compromise would involve basic usage rate calculations being checked on a sample basis using an automated device able to apply rates to events on a simple per event basis. Any discrepancies are examined manually to determine whether they are genuine or due to a limitation in the tool. Where calculations involve some form of allowance or discount in addition to simple rerating, these are checked manually on a sample basis.

To calculate a charge for an event performed by a customer, it is of course necessary to know something about how that customer is charged. To reperform, it is hence necessary to either

- Look up the relevant customer data
- Somehow know in advance the relevant facts about that customer

Knowledge in advance is possible where checks take place on predetermined test accounts. There is a risk where looking up customer data that the data being looked up is already incorrect, hence meaning the error in the actual calculation is being replicated. The risk can be mitigated by taking data from a different source to that used for the real calculation. This source should not be populated downstream from that used for the real calculation, or the same problem of replicating error may occur. An example of a suitable solution would be to interrogate any system that records the customer's order (including changes of tariffs) instead of the main Customer Care and Billing (CC&B) database.

Though tariff data and the code of any rating or billing system algorithms are separable, they are both necessary to completely describe how to calculate the charge for any given event or customer. The separation is somewhat artificial; tariff data enables flexibility, but potentially at the cost of processing efficiency. As such, there is no definitive split between how much a complete calculation is described in the code of a system and how much in the tariff data. On this same basis, there is no definitive split for any automated system doing reperformance. However, a good principle is to try to get a similar balance in the reperformance system as found in the system(s) used for actual charging. Of course, there is no such logical distinction where calculations are being reperformed manually.

Selecting Samples

It is not necessary to test each individual meter to form an opinion about meter performance in general. Testing every meter may be onerous because of the cost of implementing relevant equipment to enable testing to take place. This may be exacerbated by the geographical distribution of meters. To manage costs, the key question is to determine a representative sample of meters. A sample is representative if, for the properties being tested, the sampled meters behave the same as all other meters in the population. For example, if two devices are known to be identical in all regards relevant for their function, it is possible to draw inferences about the operation of both from the results of testing one. However, if they are fundamentally different, it would be inappropriate to base a conclusion on one using data obtained from testing the other.

If you start with entire population of meters, the question can be described as to whether

- The entire population is homogeneous; which means that each meter can be considered identical with respect to the operations being measured
- The population is heterogeneous and needs to be segmented into a number of different subsets, where each subset is homogeneous

It is only valid to draw inferences on meters homogeneous to any meters that have had their performance measured. So if the aim is to measure performance across a range of different types or builds of meters, it is necessary to measure performance of at least one meter in each subpopulation. However, the total number of devices that perform metering is likely to be small. If consistency is established, it is not necessary to test more than one meter in each subpopulation.

Populations can be shown to be homogeneous through controls over consistency. For most meters, the principles to be established, and the controls to show that they are homogeneous, would be as follows:

- Consistent hardware, shown through the same specification
- Consistent software, shown through controls over the installation of software to the production environment
- Consistent configuration and operational settings, including any data input as part of configuration, shown through controls over the implementation and testing of new configurations and both automated and manual consistency checks between configurations

It is of course easier and more beneficial to implement controls to ensure consistency rather than independently measure each meter as if inconsistency between the meters was expected.

In any situation where sampling is performed, there is always a risk that the wrong inference is drawn about the integrity of charging because the sample did not fairly represent the population. This can be mitigated. There are two general techniques to confirm and rectify sampling error:

1. Increasing sample sizes
2. Extrapolation by forming mathematical connections between known quantified properties for the population and the property that is being tested by the sample

Increasing the sample size is the safe, simple, and conclusive means of determining if the original sample suffered from sampling error. However, this may be expensive. It may sometimes be quicker to infer sampling error by observing that other properties, related to the property being measured, are not represented in the right proportion relative to the population. An example of the latter would be concluding that a certain error in calculating discounts is overrepresented because the sample tested overrepresents the number of customers entitled to that discount. In the latter case, it may be possible to adjust results mathematically, without further testing. For example, divide the number of discount errors by the fraction that reflects the ratio between the number of customers entitled to the discount in the population and the number of customers entitled to the discount in the sample.

Determining a Test and Measure Plan

For capture, the test plan will focus on establishing the completeness, validity, and accuracy of the records it generates. Completeness is that a record is generated for every sale. It is measured by checking whether a known sale resulted in a record of that sale. Validity is that every record created reflects a genuine sale. It is measured by checking whether a recorded sale can be substantiated by evidence of a real sale. Accuracy is the concept that the record generated of an event correctly describes that event. What needs to be recorded, and hence the accuracy expectations, depends on how the charge will be calculated. Here are some common examples of the information that needs to be recorded:

- A party
- B party
- Event type
- Start time
- Duration

Conveyance is checked by ensuring that all meter output is correctly presented to a bill. The check is, in effect, a simple reconciliation between the two. This reconciliation may be performed by

- Reconciling a selection of records two ways, from meter to bill and bill to meter
- A one-way check that a selection of records per the meter are represented on the bill/customer's account, and a separate one-way check that a different selection of records per the bill/customer's account are recorded as at the meter

Performing two separate one-way checks may attain the same goal as performing a single two-way check. The choice of whether to implement a two-way check or two one-way checks depends on convenience, as they are theoretically equivalent.

However, implementing two separate one-way checks does have one major advantage in terms of method. Implementing two one-way checks eliminates any risk of bias in selecting the records to be checked. Bias may occur in a two-way check if the records are exclusively selected from one of the two points being reconciled. For example, if all the records selected for a two-way check are taken from the bill, then there is no possibility of finding an error of incompleteness, because all the records are already known to have propagated from meter to bill.

Calculation is measured by reperformance. The issues in reperforming calculations are described above. Because of the difficulty in measuring calculation integrity, communication providers frequently reperform only a small sample of calculations. They instead tend to focus resources on preventative controls to ensure calculations are performed correctly in the first place. As an alternative, they may measure the accuracy of the reference data used for calculation in isolation. This can be done by adapting the same testing approach as used for capture and conveyance of chargeable events, but with the subject matter being capture of tariff changes as proposed by the marketing function, and "conveyance" of this to what is actually implemented in the system responsible for performing calculations.

Measuring the Fourth "C": Cash Collection

Eric Priezkalns

The Three "C"s of capture, conveyance, and calculation are a useful way to analyze the processing steps from the supply of a service to presenting a charge on the customer's bill or account. However, many revenue assurance (RA) functions define their scope to go beyond presentation of charges to include even the receipt of cash from the customer. We can extend the Three "C"s to include a fourth C—collection—that can also be measured and can contribute to the same methodical approach to evaluating the integrity of charging and settlement.

Unlike the other "C"s, the measurement of collection is relatively straightforward. There are far fewer epistemological challenges. The data is easier to manage. It is also easier to identify if there is missing data. The value of the services charged is known (if not, the issue is with one or more of the first Three "C"s). Cash received from customers, as long as allocated to the right customer account, is also known. The amount owed by the customer, or, in the case of prepaid, the credit built up by the customer, is the difference between the two. Hence, the "measurement" of conversion may seem utterly trivial. The emphasis is rather on timeliness of cash receipt. However, sometimes the account balance will be updated without any receipt of cash. An RA team will typically want to take special notice of these noncash transactions:

- Goodwill and other service credits
- Bad debt write-offs

Either of these may be indicative of serious leakage problems, including internal as well as external fraud. Some functions will intentionally categorize the value of credits and bad debt write-off as leakage, even though they are unlike other forms of leakage in that they may not be wholly in the control of the business. Furthermore, it may not be possible nor desirable to reduce these leakages to zero. For example, even when the customer is in the wrong, it may be better to give them a credit and to retain their ongoing value as a customer. Bad debt is always undesirable, but no system of screening customers will entirely eliminate bad debt.

Transition from the Indicator to the Director

Ashwin Menon

The last 7 months had me working with a team on optimizing the approach to revenue assurance (RA) in terms of "indicating" to the analyst the key areas for improvement. The TM Forum already has quite a good set of key performance indicators (KPIs), but as I have mentioned before, some of the communication providers in the Asia-Pacific region of the world have highly specific areas for monitoring, some of which are far too atomic to apply the TM Forum's KPIs to.

The problem that lay before us was to setup a framework and not a point indicator to the leakage. Toward this end, the shining knight said "tally ho" and set forth to discover the "holy grail" of RA: the key performance vector (KPV). As I am sure most of you know, a vector has two components—magnitude and direction. What I wanted to do was try and see if the existing set of KPIs could be converted into KPVs. At this point, via the magic of intuition, I can see quite a few eyebrows going up and a few moustaches being twirled at the thought of "yet another metric?!?! RA loves the metric system more than the UK does."

The reason behind the KPV is what I would call a goal cascade (GC). The intent of the GC is to enable the various subsections of the RA team to be aligned with the highest-level RA goal of the organization. For example, when a communication provider says, "0.4% leakage is acceptable to me—anything beyond it calls for discussions with HR," how does the analyst know the weight of his/her function to that 0.4%? Furthermore, how should he/she interpret the functional KPI? The KPI definitely does highlight the true state of affairs at that moment, but is it in need of further improvement or should the analyst call his mates and nip down to the local pub for a pat on the back?

The GC is an engaged activity where the trickling of the key goal to the various subsections of the RA team is cohesive—and trended over a period of time. Now comes the kicker. The KPV is not an "in-point-of-time" indicator. The KPV is the second derivative of the KPI magnitude trended over a period of time (it is not *exactly* the second derivative, but that is the closest generalization). The KPV actually involves quite a complex formula taking into account multiple factors. If you hate mathematics as much as I do, you have probably fallen asleep at this point. However, on waking, the end result of a KPV is an in-line representation of

KPV = magnitude (from the KPI) + functional performance up/down
+ alignment percentage to key goal

The KPV is not meant to replace the KPI. The KPV is an extension of the KPI itself. The KPI builds structure in a vast science (Einstein used to wonder about two things—the boundary of the universe and second, the boundary of RA—of course I am joking). KPVs come in useful when the RA team requires a synchronized view of KPI aggregation.

The KPV, in absolutely simple terms, is an up/down indicator (which is derived from past performance) and a measure of how much that up/down change will positively/negatively impact the organization (derived from the key goal cascading). I know that the concept itself is quite a bit to read and understand immediately (I have been working on this for months on end, and it still is not completed). Let me see if I can break this down further:

(a) GC refers to how a high-level goal is broken down into smaller and specific goals for each subsection of the function. For example, an RA goal is broken into subgoals for the prepaid team and postpaid team.
(b) The subsectional goals define the outer limit of negative performance, which is continually updated based on KPI tracking.
(c) The KPV is created over an existing KPI by deriving it from the historical performance trend. This is a bit more complex than a simple averaging over time.
(d) The KPV now indicates the performance of functional departments in conjunction with the key goal.

Worked Example

Let us assume that communication provider XYZ has visible revenue of 100. We assume this is the revenue at risk.

XYZ have said that 0.4% leakage is the threshold. This translates to 0.4% of 100 = 0.4 units.

Now let us assume XYZ has only two divisions, for the sake of simplicity. These would be the prepaid division and the postpaid division. The stake of these divisions is divided as follows:

Prepaid—85%
Postpaid—15%

How do we derive the stake? Well, it is a mix of the subscriber base split, the billed records/MOU, product spread across these headings. Now, back to the scenario.

Prepaid threshold is $0.4 * 85\% = 0.34$
Postpaid threshold is $0.4 * 15\% = 0.06$

The aforementioned is a very simplistic representation of a GC. It goes down further into the individual KPIs, for example,

In prepaid, 0.34 = subscriber misalignment (SM) KPI (assume it constitutes 15%) + records lost (RL) KPI (assume it constitutes 85%)

This means SM = 0.051 (15% of 0.34)
RL = 0.289 (85% of 0.34)

Now the highest thresholds for each region are defined in the KPI. Moving forward, giving an example of the KPV is extremely difficult, but 2 weeks down the line, a KPV reading for RL might read like this:

RL-KPV = 0.141 m (representing the magnitude) || <> (assuming performance has improved over the last week) || 49% under recommended threshold

Performance Management

Güera Romo

Many years ago, I was contracted to do process reengineering at a government department. Somewhere during this time I became involved in an ISO certification exercise for this particular department. Not having had much experienced with ISO implementations, I asked the project manager to explain to me in brief terms what ISO quality standards were. He responded with, "document what you do and do what you have documented."

That has pretty much become my motto for revenue assurance (RA), and much broader really. This pertains to anything that should be documented. Business rules, business processes, system flows, system rules, you name it. Very often these are not documented, which means we all do what we think we should, or are capable of doing given the uncoordinated chaos between systems and departments that is most often the norm.

It becomes difficult to assign and monitor key performance indicators (KPIs) because you are not working from a known and agreed factor. Much like yellowware: firm enough to touch but not solid enough to hold. I have seen many first stabs at KPIs and these are based on gut feel or general common sense aspects we should measure. These KPIs were not scientifically determined or based on any maturity index of the department's capability and/or capacity.

I have recently spoken to a number of junior-level staff, both in the RA field and other industries and disciplines, on the topic of defining KPIs for their business functions (such as procurement and service level agreement [SLA] management) and their individual performance (how are they performance managed against the job's KPIs). I was astounded to find that both the human resources people I spoke with (one of whom was senior) did not see the relevance of the continuous string of interconnection between the organization structure to the job role (stating the role objective in context of the strategic and tactical plans), to defining the job outputs (those core responsibilities that would add the execution view to the strategy), to the KPIs to measure and manage the contribution to the organization. A balanced scorecard was some academic thing that resides with the Head of Department and does not filter through to the job description. That means the performance management chain is broken.

I also found that all KPIs discussed with me assumed a process maturity of between 3 and 4. In other words, it assumed the processes involved in producing the output to be measured, or support processes to enable the output, are all in defined and managed mode. I did not find one KPI that was aimed at establishing a capability, as you would assign for a level 1 or 2 maturity. Needless to say, those poor individuals with these sky-high targets did not have the basic doing capability in place let alone the measuring and reporting capability. For some reason organizations assume that the infrastructure and processes needed to run and grow the business must already be in place or else are relatively easy to just do as part of the overall job.

It was quite alarming to realize how few individuals thought their personal KPIs had any relevance to any measurements that might be in place for specific financial or service-related work they may be involved in. This I would put down to not speaking to an individual who had both a set of KPIs used for annual performance management as well as measurable and reportable work tasks, typically the statistics you would find in a call centre, help desk or RA. The customer service and RA people I spoke with did not have a Balanced Scorecard and those individuals with a balanced scorecard do not work in an environment where objective performance measures were taken.

I have seen this in a communication provider and have tested it now at two banks. They all suffered the same mindset. There is a disconnection somewhere, a fragmented view on this complex whole we call an organization. Yet, come the financial year-end and we are back to drawing up the new balanced scorecard because last year's failed.

Am I seeing a connection where there is none, or have I just not seen this implemented anywhere?

Benchmarking

Güera Romo

Mark Graham Brown wrote a succinct little business-savvy book called *Get It, Set It, Move It, Prove It: 60 Ways to Get Real Results in Your Organization* (Brown 2004).

The book contains a number of short chapters neatly summarizing basic business issues or mistakes made by organizations in its attempt to achieve its objectives.

One such chapter is on the use of benchmarking as a way of obtaining ideas for business improvement. Tongue in cheek, he refers to a benchmarking trip as an opportunity to go somewhere warm and fun, while you tell your boss you are doing a benchmarking study.

The idea of a benchmarking study is that you get some great ideas for improving processes by studying other companies that perform with much greater efficiency or at a lower cost than you do. Benchmarking should then be a way of shortcutting process improvements by letting other companies do the trial and error until they hit on an approach that works.

While I read this book I looked for the application to revenue assurance (RA). Brown lists a few shortcomings of benchmarking that I believe are equally valid in our industry. Think about these common shortcomings in the context of furthering the RA standardization effort:

- Well-run businesses are inundated with requests to do benchmarking, so much so that one particular car manufacturer started asking money for benchmarking tours and ended up making more money from the tours than from manufacturing cars.
- Many companies who add themselves to the benchmarking databases are a legend only in their own minds. They volunteer information to others but are actually so outdated and generic that it is a total waste of time to review what they have to offer.
- Benchmarking often lacks focus and preparation. Benchmarking should focus on a singular goal and process. Once the process has been identified, a plan for selecting the comparison data should be done. The outcome of an unplanned benchmarking study would not contribute to the organizational learning as you would realize after the field trip that you missed vital information or may even have selected an inappropriate partner against which to benchmark.
- Thinking that you have to benchmark against a large and well-known company instead of a small company that actually does things differently is another common shortcoming of benchmarking.

When I read benchmarking reports I do not always understand what was benchmarked. I would think that we would like to take X and compare it against a number of sources to determine if X looks and feels the same when compared to others. I can see from the benchmarking questions that X is implied but it is not explicitly stated or described/labeled as X.

What is the expected benefit of being benchmarked? Many companies are looking for the perfect role model against which to benchmark themselves in the hope that they would learn how to do the job. Very often they do not have an idea what should be done and how. The exercise is seen more as a hunt for free advice, templates, and "how to" knowledge. If they find a company against which to compare themselves, what is the benefit to the comparison company and how many companies end up training or guiding the benchmarking company? Is this still considered benchmarking?

What Happened to the Subex–Analysys Survey?

Eric Priezkalns

Some time ago, you could find a curious anomaly on the home page of Subex, the Indian software vendor. If you looked at the centre bottom of the home page, you would see a link that was no longer a link. It read "Operator Attitudes to Revenue Management Survey," and if you dug into the HTML source, it is obvious there used to be a link to another webpage, but that it had been stripped out. It turns out the most famous and longest-running survey in the history of revenue assurance (RA) was last performed in 2007 and has not been performed since. Why did it stop?

The year 2007 was the fifth year running when Subex had engaged the research firm Analysys to survey the estimates of leakage from around the world. The survey had been a tremendous success in garnering press coverage and molding people's ideas about what RA is and the scale of challenge it faced. This survey, with its seemingly always-growing X% estimate of leakage, was guilty of spawning opening paragraphs for a few hundred thousand consulting reports, all of which began the same way:

> Average revenue loss in telecoms operators is [scary number] percent, which equates to a loss of $[scary number] every year. Because of that alone, this report has been proven to be worth the $[scary number] you paid for it, even though all we did is copy an *estimate* quoted in somebody else's survey* and offer no real facts to support the claims made in the first sentence.
>
> *or maybe we just copied the opening paragraph of somebody else's report and have never even read the original survey

I joke, of course. No report contains the second sentence, though many should. If the reports had been written by lawyers I expect there would have been many more caveats about the mental leaps that got people from ill-defined estimates to absolute conviction of revenue loss in every business to justification for any and every activity designed to reduce leakage, whether suited to the needs of the communication provider or not.

I take the disappearance of the survey to be a positive sign. A leakage survey is no good unless it scares people, and there was always going to be a problem scaring people every year, year after year, without fail. After all, nobody advertises a horror movie sequel by promising fewer chills and frights than in the previous film. The problem with increasing the fear factor every year is that it leaves everybody working in ongoing RA with the headache of explaining why everything keeps getting worse, even though they are paid to make things better. After a while, the CFOs and CEOs get fatigued with frightening stories about how much money their business is losing. More important, they start asking questions about why their business was not noticeably richer by the promised billions of dollars, and noticeably more successful than its competitors as a result. Since the disappearance of the Subex–Analysys survey, the quoted estimates of other surveys have tended to become more conservative, though we still see "professional" research that aims to generate more shock value than useful data. Yet, oddly enough, the fall in global revenue leakage from the previous oft-quoted highs of 10% or even 20% to the rather more restrained levels of a percentage point or two has not been matched by a commensurate (and otherwise unexplained) rise in global operator revenues. In short, the CFOs are right to be skeptical about the scale of promises made for RA. That is not to say that RA does not have a sound business case. A 10-fold return on investment is still a perfectly good reason to put money into RA, and can be delivered by improving the bottom line by only a fraction of a percent. Less hyperbole and more precision is a sign that people have shifted from making an abstract case for RA to actually doing it. In that context, I take the disappearance of the Subex–Analysys survey to be a positive sign, although I am sure it served Subex well over the years.

Children are frightened by scary stories about unseen monsters hiding under the bed. Adults know not to take them seriously. Subex giving up its popular annual study tells us that the RA industry is maturing, or as I prefer to put it, it is growing up.

When $25 Billion Is Not Worth $25 Billion

Eric Priezkalns

In 2006, SubexAzure (as they were then known) announced that African telcos were losing a total of US$25 billion every year due to revenue leakage, which is the equivalent of a quarter of their revenues. Move forward to the present day, and you can still read doom and gloom predictions about revenue loss in developing countries. But when dealing with numbers of this scale, can the hype be substantiated?

Stop and think about that 2006 announcement for a moment. Imagine that those telcos had cut their leakages to zero in 2007. They would be US$25 billion richer, right? *No.* Of course not. Communications services are price elastic and there are not many countries left were operators are shielded from competition. That means if a communication provider charges more for its services, then people will choose to buy less services from it. In my experience, the poorer the customer, the more price elastic they are. It is nonsense to suggest that African customers have a spare US$25 billion per annum, rolling around in their pockets just waiting to be handed over to a telco.

Even if all African telcos got a lot better at recovering revenues, the likeliest impact is that they would (a) have to reduce prices or (b) suffer reduced total business. This would offset their improved recovery, which means that the financial benefit from fixing leakage would be much less than the nominal leakage figure. They would also decrease directly variable operating costs too, but these costs to one telco are revenues to another telco. In the final calculation, improved recovery to the point where customer bills go up by 10% or more is going to provoke a customer reaction and so cannot be simply assumed as a straightforward and realistic goal for improving revenues. After all, many of the customers who make the most use of telco services—fraudsters and bad debtors—have no intention of paying anyway.

Global Revenue Leakage to Fall (?!)

Eric Priezkalns

Just when you think you know the stock conclusions that will be in every report about revenue assurance, you find one that surprises you. Juniper Research's 2007 report (Dehiri 2007) concluded that global revenue losses were set to fall. Interesting. Not that I take research into revenue loss seriously. The difficulty of measuring revenue loss is high. Meanwhile, RA practitioners exhibit too great a propensity to make guesses and succumb to bias when offering numbers to researchers and the rest of us in the public domain. I have observed many a communication provider's private, internal report on leakage that would put their leakage at somewhere between 0.1% and 0. Inevitably, these numbers are not being factored into the headline-grabbing surveys. But, however, cautious we should be about this particular finding that revenue losses will fall, it does make me think of a few questions (although I admit I have also thought about the answers already):

1. Why did the Subex/Analysys revenue leakage survey and similar surveys fail to ask what revenue assurance managers think revenue loss will be *next* year as well as in the current year?
2. Would that lead to a change in results that Juniper and other research companies predict?
3. Without asking that extra question, is there any incentive for revenue assurance managers to be realistic when giving their opinions and predictions?

I think you can work out where my thinking leads …

Cost of Nada

David Leshem

I am somewhat troubled why revenue assurance (RA) attracts only limited interest from the "typical" communication provider CxO. I will try to offer a suggestion why …

The highest art of any trade is being able to answer the "what if not" challenge. I presume no one would dare to ask, "what is the cost of not have a billing system?" or "what is the cost of not having a customer service department?"

Yet, I am not so sure what would be the reply if we challenged the RA function and the related costs it involves. Sure, there are rather useful RA dashboards, and there is documentation about the right way of doing business by the TM Forum, and plenty of consulting firms. We are also very familiar with fancy return on investment figures which support the cost-benefit argument to implement an effective revenue assurance policy. However, my challenge is being able to reply to a simple question: "what is the cost of not having a revenue assurance function?"

For sure, one can craft a reply and mention Sarbanes–Oxley Act (SOX) as a supporting argument. Others would mention proper financial controls. In some cases we can offer an uphill reasoning that we need RA to demonstrate that the communication provider is taking all the proper measures to shield itself from class action lawsuits when billing is not right. To my ears, these are somewhat whining arguments. I am looking for a clear and decisive answer, similar to one where no one dares to ask the cost of not having a billing system. At least in the case of billing, the reply lays within the question.

Myth of Average Intercarrier Leakage

Eric Priezkalns

Let me paraphrase one kind of nonsense you hear said about intercarrier leakage from time to time …

> *Intercarrier revenue leakage represents approximately 2 to 7 percent of a communications provider's costs for traditional voice, and is even worse for IP.*

One of my favorite bugbears is the idea that there is an "average leakage" for intercarrier traffic. Think about it. Take every telco in the world. Estimate the net error when they pay each other for intercarrier services. What is the mean net average leakage? Zero. We know that as a fact. We know this without estimates or data or guesswork or case studies or anything. If the purpose of the estimate was to get an average across *all* telcos, and we are talking about only intercarrier traffic and settlement, then we know that one telco's loss is equal to another telco's gain. In short, errors in intercarrier billing and settlement are a zero sum game. If one telco is being charged too much, another is earning too much revenue. If one telco receives less than it should, another is paying for less than it used. Whatever gibberish you spout, there is no way to identify an average "leakage" without also admitting to an average "gain" that is equal and opposite. Yet some vendors are so addicted to marketing spiel, they cannot stop themselves. Bang goes their credibility as they try to tell us a mathematical impossibility has been proven thanks to an unscientific observation that also happens to be very convenient for vendors (if their customers are gullible enough to believe them).

Raw Data, Workflows, and Mr. RA Analyst

Ashwin Menon

I have started to notice a significant shift in the communication provider's attitude to leakage detection and correction. There used to be a time (long, long ago in a galaxy far, far away ...) when the most important item for most revenue assurance (RA) teams was a set of crisp and clean dashboards that tell them:

> Hey, Mr. RA Analyst, I'm working on dimensions and measures which have told me that you have 0.3511 percent leakage in product SuperSaver199 between the mediation and billing.

This used to suffice the needs of Mr. RA analyst. He went forth armed with "0.3511 percent," "SuperSaver199," and "Mediation vs. Billing," and reported the same to his network team. Naturally, Mr. Network Guy wants a sample set of records so that his team can go about plugging the leakage. Unfortunately, Mr. RA Analyst works with dimensions, measures, and dashboards only ... so he sets forth on another activity (a search for the leakage grail) where he tries to pull out the raw data from the network records that corroborate his claim. While he proceeds on his quest for data, the leakage continues unabated. Furthermore, his view is a bit myopic as the leakage might have been analyzed from a data transfer angle, but might not have been investigated from an "impact to customer" angle.

What I was trying to illustrate in this fictional anecdote, but drawn from real-life experience, is the need for raw data and established workflows. I am a great believer in getting down to the raw data and validating the actual data flows between interworking systems and validating against expected business process flow. Recently, I have been interacting with operators who share the same view about working with raw data, as opposed to running an RA department based on pretty dashboards and metrics from data warehouses.

There is a visible paradigm shift in the way that operators are setting up RA processes in the Asia-Pacific side of the world. The business process of leakage reporting has matured significantly, where every issue needs to be reported with the associated "proof from network," which usually constitutes the discrepant/mismatched records. The network and IT teams even tell the RA team what particular fields they expect in the report for their cross-verification. It is important to have leakage reporting AND tracking. We are now seeing cross-functional teams that have been given the task of ensuring that issues raised are being closed in a timely manner. When I say cross-functional, I am talking about workflows that involve a first level of investigation by the RA team, impact analysis by a finance function, technical cross-verification by IT, corrective action by a network task force, and rectification corroboration by RA and IT. I like the approach primarily because there is a shift in the way RA is now viewed as more than just a "finger-pointer."

The emphasis on being able to "drill down" to the raw data is something critical to the success of a revenue assurance function, because here we are questioning the fundamentals of the underlying data, its behavior in various complex network systems, the impact from a technical/financial/customer/service angle, and so on. The natural evolution of a multi-impact view of leakage analysis is growth from a mere system health-check validation to something like genuinely customer-centric revenue assurance.

On Analytics, Dashboards, and Revenue Assurance

Mike Willett

From my perspective, there is little doubt that detailed analytics are essential to be undertaken by revenue assurance (RA) teams and should be the primary activity. Dashboards serve some value in identifying trends but I suggest that any movements upward or downward in trend lines (such as the 0.3511% in Ashwin's article entitled *Raw Data, Workflows, and Mr. RA Analyst*) become almost impossible to justify as leakage. Let me explain my views.

First—what is the right baseline value to begin a dashboard with? Assume every day for the last 3 years, 100 calls come into a mediation device and 70 come out. So you set a baseline at 70 and if this drops to 65 then you decide to follow up with the mediation people. Great theory, but who says 70 is correct in the first place? It may well be that there is a leakage in the 70 and in fact it should have been 80 all this time. The only way to know where to set a possible baseline then is by doing a detailed analytic. So let us now assume we did that and we know it should have been 80, we have fixed the mediation device and we start seeing 80 go through every day. Great stuff: we found 10 lost calls, fixed that issue (hopefully got a pat on the back), and are now monitoring every day just in case, this drops back to 75 in which case we will be on the phone and get that problem fixed as well. Two weeks later, and the volumes hit 70 and they stay there for the next few days. You are on the phone saying something is wrong, no one believes you and ask you for the call detail to prove it. You do the analytic again and find that 70 is the right volume as new products have been introduced and the mediation device treats them differently, or there was more of a particular call type in that time period due to a marketing promotion, or … etc. My point is that the reasons could be endless and setting a good baseline dashboard value would only be possible in a stagnant, never changing, environment—not really the kind of environment you find in most communication providers. And if every time you called out "leakage!!!" when there was some other change you would be (1) busy trying to prove this and (2) lose your reputation for integrity and the time of the people who you keep raising alarms to.

Second—many revenue leakages reside in the less than 1% bracket. Sure, there are some massive ones and these are the ones that we talk about at conferences or hear of in training but the normal reality is that most are so small, relative to the revenue pool, that a dashboard would find impossible to discern them from other activity (similar to my point earlier). But if you have a $100 million product, then finding $1 million is a good result but to find it is likely to need detailed analytics.

Third—only with some detailed data can you go back to the business area and tell them what specifically needs addressing. Analysis may allow RA to determine what to fix specifically. Analysis can also tell you the extent of the issue, making quantification of the total impact more easily achieved, and this can help drive prioritization of resources to address this, relative to all the many other competing demands across the communication provider for money and people's time.

Lastly, there is a very high cost and lots of effort needed to maintain an RA dashboard that adapts for all the business rules and all the changes that take place in a communication provider. In fact, I would contend that it would be almost as high as putting in the operational system changes that it is meant to monitor. Why? To keep it current with respect to all the network and IT changes going on would not be insignificant. Add in all pricing changes, campaigns, what the competition is doing, and how to account for the behavior of people, and it is practically impossible to model it all. To make a point, I ask myself how many times I have heard stories of RA managers raising alarms at a 50% drop in traffic over a 2-hour period, only to later realize that this was due to the nation being absorbed by a football match (and the RA manager probably was watching the TV as well).

Do I think there is value in RA dashboards then? Well, yes, but they have to have some key attributes:

(1) Data presented has to be summarized correctly from accurate underlying data (avoid garbage in—garbage out).
(2) "Alerts" have to drive an action (e.g., doing an analytic, raise a task for the billing team).
(3) More true-positive alerts are generated than false-positive alerts (the challenge faced by our fraud management friends as well).
(4) Data needs to be delivered in a timely manner (there is no point finding a leakage after everyone else in the business has).

Putting all these into play requires time, thought, and effort. And that effort could be spent on doing a detailed analytic. Getting the balance right is the challenge!!!

Progress Key Performance Indicators

Ashwin Menon

Now here is a question: can revenue assurance practices be standardized (or bottled) into a one-size-fits-all solution?

What do I mean by the aforementioned statement? Should it be imperative that every issue/leakage raised by the analysis of underlying data yield a "bigger-picture" analysis, or should each issue raised be closed on a case-by-case basis? It is still a fairly common practice to reconcile data on atomic levels (xDR levels) by performing transference checks through all the downstream systems. In effect, we could say that we are going to revalidate the expected business process defined for a particular product. However, is the effort and time that we would need to put into this exercise worth it? I am not of the opinion that we do not need to validate transference, but do we need to be matching each and every xDR and check for its presence in the downstream systems? The number of analysts required to crunch the outputs of such large data correlation check and the time that would be going into it would not be, in my opinion, a correct and efficient use of the resources at your disposal.

If we go by the key performance indicators (KPIs) defined in the *GB941 Revenue Assurance Guidebook* from the TM Forum (Priezkalns 2009), we could see a fairly good set of indicators for monitoring the overall health of a system. However, it is my opinion that it has been formulated from the viewpoint of an operator with a good level of maturity in RA, with emphasis on a process-driven methodology. Would this approach be accepted by every operator who wants to set up an RA department? In my personal opinion, I do not think they would. Over a period of time, once the initial steps have been taken, then perhaps the operator would gravitate toward a more standardized approach, but in the initial few months everyone runs around trying to justify the RA function. This is essentially a period of "controlled chaos" where we are looking for all the quick-win scenarios. Unfortunately, the emphasis is only on discovery and not recording the methodology for ascertaining best processes toward progressive maturity. Even though this is a period of individual heroics, it could lead to the development of a viable process by the various learning that one gains.

Using Analytics to Enhance Sales and Marketing

Eric Priezkalns

Analytics cannot tell you how people in general will behave in the future, but it does tell you how your customers have behaved in the past. Businesses spend a great deal of money on market research. Even so, many mistakes are made, not least because of the risks of sampling error, of asking the wrong questions, because people give biased answers or because there will be factors in real life that may not be anticipated in the design of the research. In contrast, the current customer base will often provide a larger population than any that could possibly be assembled through another kind of research exercise, and will also permit insights to be drawn over a much longer period time. Current customers are an authentic representation of real life. To effectively exploit this valuable resource of information, you need two things. First you need the ability to collect, store, mine, and analyze the data. Then you need to know what questions to ask.

Collecting, storing, mining, and analyzing data. Does that sound familiar? Sounds like the kind of thing you need to do for revenue assurance. There will be strong feelings about whether revenue assurance people should concern themselves with assisting sales and marketing. I myself made the argument that revenue assurance works *"without influencing demand"* in the TM Forum's definition of revenue assurance. The point of that statement was to exclude the promotion of increased customer sales as an element of revenue assurance proper. However, whether or not it is revenue assurance, it will offer an opportunity to revenue assurance people to get increased value from the tools, skills, and data they already possess. If they can do it, and if they are finding that returns from addressing leakage are diminishing as a result of their hard work over time, is that a bad thing?

A business may possess a lot of data, but data is not the same as information. For data to become information, there has to be a use. Knowing that families have 2.6 children on average is not useful, because no family has 2.6 children. By the same token, inferences about "most" people, the "average," and all that are probably just convenient fictions. They make it easier to think of how large numbers behave, without really understanding how each and everyone in that large number is actually behaving. Successful analytics involves drawing conclusions from actual individuals, by cutting the data based on theories of how to group individuals into coherent groups that tend to behave the same way. This makes marketing both personal and relevant.

One of the most effective uses of analytics is to tailor prices and promotions to drive increased sales and improved margins. Provide a selection of customers with a better rate than normal, and see if the increase in sales justifies the lower price. The connection between data and customer is already established for postpay customers. The rise of mixed postpay/prepay plans or offering enhanced services and special offers via the Internet can be ways to increase the span of knowledge to prepay customers. Future promotions can be targeted at the customers that are most responsive. This will enable the business to segment its customer base according to the different utility curves of different customers, and hence refine its pricing strategy and offerings accordingly.

Analytics can be used to assess the relative profitability of competitor's pricing schemes and incentive programs, or to perform hypothetical what-if analyses of proposed new prices and incentives. Particularly where there is highly stratified pricing, with rates decreasing as consumption increases, comparing new prices and pricing points to current customer behavior will help to understand the potential for revenue growth or cannibalization depending on the extent to which customers opt to buy more or less. Overlaying the volumes of sales per individual customers with competitor's pricing will identify which customers would be better off if they switched to competitors, and which customers are benefiting from the best available deal at present. This may help with identifying price reductions that would draw price sensitive customers away from competitors, and also price rises that would still better the offers made by rivals.

Adding data on cost of sales changes the focus from revenues and puts it on margins and profits. In a similar way, it is possible to include data on the timing of cash flows, say from bulk purchases of "bolt-on" usage allowances, to gauge how these can be improved through understanding the segmentation of the customer base. Any numerical data on costs and revenues that can be associated with individual products/services and with individual customers can provide a rich basis for comparison to competitors and preassessing the impact of proposed changes to prices and offers.

Customers that superficially seem profitable may be viewed differently once all costs are taken into consideration. Time spent handling customer complaints, or a track record of returning goods, may indicate the customer is more costly to serve and less desirable than originally thought. When identifying which customers to offer loyalty benefits to, it is worth directing these benefits to customers that are cheapest to serve by virtue of the smaller demands they place on the business. For example, there may be incentives for customers to purchase online because of the lower cost of taking the order compared to processing a sales order through a call centre or in a store. It makes sense to extend that logic by prioritizing customers that place a smaller burden on the business, for example, because they submit orders in a way that consumes less staff time or because they raise fewer customer service queries.

Revenues and costs are readily susceptible to analysis because they are numeric and because it is relatively simple to gather the necessary data and associate it with products and customers. In practice there are many ways to measure and segment customer behavior, and hence look for trends and groupings within the customer base. Geography may be a factor in sales and costs, thus enabling different strategies for different locales. Geography can be analyzed by either the customer's home address or, in the case of mobile phones, where the customer is when the phone is switched on. Demographic factors like age or cultural leanings may also provide a viable basis for analysis and segmentation. This kind of data might be obtained via credit checks or by asking customers to submit to a survey.

What data will be relevant depends entirely on the product on offer and the nature of human behavior. Generalization from past experience may be useful, but will be misleading if nobody has considered some options for altering the offering or segmenting the customer base. No amount of data will assist in drawing useful conclusions if the wrong questions are asked. That takes the skill of the imaginative marketeer. But harness that insight to data and the power to perform analysis, and, like a scientist, the skilled marketeer can progress from forming a theory to being able to corroborate it in practice. This will lead to better decisions made with greater confidence and improved understanding of the results that are subsequently achieved. Imagination combined with data makes for good business sense.

Double-Edged Saw

Eric Priezkalns

If you hang around revenue assurance departments long enough, you notice a curious thing. Most of the time they report that losses are going down, because they keep finding things that are wrong and keep fixing them. But ever so often something remarkable happens. And it usually happens at just the point in time when it seems that everything has been fixed and there are no more leakages to plug. The description may vary, but essentially a "new" kind of loss is suddenly discovered, or the scope of revenue assurance (RA) suddenly grows, and the value of reported losses shoots up. Of course, the RA team gets back to work and starts to diligently reduce the losses again until they nearly reach zero when ... they suddenly shoot up again. So a graph of revenue losses against time ends up looking like the jagged edge of a saw blade.

Of course, somebody cynical would say that the losses fall over the year to show that targets are being met. And somebody cynical would say that they shoot up at exactly the time it comes to set budgets and targets for next year. That cynical person would say that the numbers reported are more to do with securing ongoing headcount and investment in technology than they are to do with the actual level of benefits delivered. But I would not say those things ... I would say it is often to do with people changing jobs. If the guy in charge of revenue assurance changes, the old boss wants to leave boasting that everything is fixed, while the new boss wants a report that justifies his budget—not one that justifies cutting his budget. By the same logic, if there is a change in the executive who reads the reports (yes, it is rare that an executive reads RA reports, but it does happen) you can safely assume the business case for revenue assurance will be "bolstered" by finding a few extra causes of loss not reported to the previous executive.

Of course, the saw can cut both ways. If nobody changes job for a while, depending on simplistic reports of benefits will in the end undermine the value and purpose of revenue assurance. And, incredible but true, but some executives are smart enough to see through the implications of a saw tooth report of leakage. So if you see a saw tooth trend in leakage reports, you learn more about the rate of personnel turnover than you do about the losses suffered or benefits added. Companies that want to consistently have the best results are better advised to work steadily through all issues, not just handle a few at a time only to be repeatedly surprised by "new" leakages. Because, of course, the leakages are very rarely new, they are only newly reported. In the cases where the leakages are genuinely new, they are usually small, because blockbuster revenues from totally new products are rare and that means leakages from new products are small in absolute terms, even if they represent a proportionately high share of the revenues generated by that new product. All of this means that any very large "new" leakages are in fact large old leakages that have been ignored for a long time—and costing the business money for all that time too.

Anyhow, this kind of thing goes on in communication providers around the world, again and again, and it is not for me to criticize if the revenue assurance team finds saw tooth reporting to be the best way to get buy-in from their business. But if you ask me, it will only be when the saw tooth turns into a flat line that revenue assurance can be said to have really delivered.

Chapter 6

Hammer and Tongs: The Tools of Revenue Assurance

Tools Overview

Eric Priezkalns

It is difficult to get a good, honest overview of revenue assurance (RA) tools. Vendors tell you about the tools they offer, not the tools they do not offer. Have you ever noticed how every year their product offerings get broader and broader, yet they never said there were gaps in the offering the year before? Consultants always play safe and tell you about the state of tools 3 years ago, so communication providers rarely try something new (though they also reserve the right to blame the consultants when whatever they buy fails to work as expected, or worse does not deliver the anticipated and promised financial benefits). And nobody talks about the niche tools provided by niche providers (except the niche providers, of course) although these may be better suited to the needs of some customers than the multipurpose offerings pushed by more mainstream suppliers. Well, if you want some advice on how to go about negotiating the minefield of automating RA, here is my modest attempt to map a way through.

This article will walk through the main issues in selecting the right tools for the job. It gives a brief discussion of how to use the 4 "C"s of capture-convey-calculate-collect to define systematically what you are trying to accomplish with a detection tool, and talks about how to match the detection tools to the objective. The article also discusses what typically goes wrong when acquiring new tools and finishes by listing generic strengths and weaknesses for different types of tool.

Two Types of Tools

Before we go any further, let us deal with the most common mistake of all when talking about RA tools. There is more than one possible goal for automation. The two possible goals for automation are as follows: improved detection (what most vendors talk about), and improved correction (which most vendors conveniently forget to mention). If you only detect, and never correct, you are unlikely to add much value to the business. However, detection is the challenge most easily solved by automation, which is one real reason why it sometimes gets disproportionate attention.

Most tools focus on detecting anomalies by comparing data sets and revealing unexpected discrepancies. However, not every discrepancy is evidence of a leak. In fact, low-quality data or poor design of the RA tool can lead to lots of false positives, where RA staff spend most time confirming that the "issues" they identify are not really issues at all. Then again, when real issues are found, they need to be acted upon. The RA team cannot simply find leakages and then have no ideas what to do about them, but in truth this sometimes happens. In the worst case, real RA departments may regularly produce reports that nobody even reads. To complete the circle, it is possible to automate workflow and incident management too. This aids the process of recovering leaks and stopping them from happening again.

Vendors have steadily made their products more sophisticated, and now often include some incident management capability with their detective tools. However, such functionality may be limited and tends only to be exercised by a small group of users. Instead of using software that is specific to RA to manage issues to resolution, it may be better to take the outputs of detective systems and integrate with existing incident management software used across the communication provider. This may initially be more challenging from a technical perspective, but technology is there to help people do their jobs, and solving the problem of communicating issues has the advantage of once and for all reducing the enormous burden in requiring people to chase other people to execute RA improvements. It also reduces the burden in reporting on who is and who is not responding to a clearly prioritized issue list, and hence automation of issue management can play a vital role in teaching the business new, and better, habits.

What Can Be Detected

There are four kinds of processing when handling any transaction with the customer (supplying a call, accepting an order for a new line, taking and supplying a request for a new feature, etc.):

1. Capture: The relevant data relating to the initial transaction
2. Convey: The data to, from, and between systems, people, paper records, and so on
3. Calculate: The monetary price for the service provided to the customer who received it
4. Collect: The money from the customer

These can be summarized as the 4 "C"s. Automated detection will aim to identify failures in one or more of these processing stages. Automation is most beneficial when performing a high volume of repetitive tasks relating to the integrity of one or more of the 4 "C"s. Automation is far less beneficial for irregular checks, as might be pertinent for activities like change control.

Detecting Conveyance Errors

I will discuss the 4 "C"s out of order to highlight what RA departments most frequently check (and what they fail to check). Conveyance errors are the most common kind of error to be subjected to automated detection. It is a simple check to automate as it involves reconciling data from one place in a data flow to another place in a data flow. These represent some of the best-known checks in RA. They are so common they have well-recognized names like switch-to-bill, network-to-bill, and order-to-provision. Because the matching logic is usually not that complicated, conveyance checks are an obvious target for automation. When processing large volumes of data where there are inherent timing differences in what data is processed by which system and when, it is easy for some data to fall undetected into error or suspense. A robust automated tool will highlight when this happens by indicating the age of the records without appropriate matches. Some care is needed to apply common sense rules. It is necessary to distinguish potential problems from the mere consequences of the fact that there is always a time lag in processing. It takes time for records to be processed so there may always be records found at one stage of processing that are yet to progress to a later stage of processing. Nevertheless, rules for automating a routine reconciliation are relatively straightforward and they change infrequently, making conveyance the easiest of the processing checks to automate efficiently.

The automated conveyance check is the biggest driver of what has emerged as the archetypal RA system: the big database designed to perform reconciliations, or as I call it, the "RAdb." The fundamentals of such a system are as follows:

- Take a lot of data feeds
- Pipe them into the same database
- Perform queries
- Look for mismatches or signs of problems

The data itself can be checked at various levels of granularity. For example:

◼ Trend analysis: compare totals to expectations over time
◼ High-level reconciliations: reconcile at the level of files, customer segment, and so on
◼ Low-level reconciliations: reconcile xDRs, individual orders, and so on

In turn, the RAdb model leads to four kinds of automation solution:

1. Acquire specialized commercial off-the-shelf (COTS) software intended for particular RA reconciliations such as network-to-bill; this tends to be the core product offered by the best-known niche RA vendors.
2. Build reconciliation capabilities into the OSS/BSS architecture—for example, unique identification tags for transactions and sequence tracking; this is an increasingly common value-add from OSS/BSS vendors.
3. Develop point-to-point reconciliations in house; this may be the quickest way to deliver tangible results.
4. Lever an existing enterprise data warehouse; with this approach, you use the existing infrastructure, augmented by new data feeds and queries where necessary.

Detecting Capture Errors

Checking capture errors involves far more than simply performing reconciliation between two extant data sets. The challenge is to find a way to assure that the initial data capture is complete, accurate, and valid. This is problematic because:

◼ There may not be an existing source of reliable independent data for comparison
◼ One mistake may go undetected for a long time (e.g., an incorrect configuration rule)

Capture errors are potentially very serious because they may be repeated very many times without being detected. Such mistakes may result from the way the staff is incentivized. For example, if staff are motivated to deliver a new service on time but without adequate concern for ensuring whether the data is complete, the service could go live without the ability to bill. Another example is the case where there are bonuses to staff based on the numbers of orders they submit but there is no quality control over information recorded in the order.

The techniques for identifying capture errors are as follows:

1. Invasive testing/mystery shopping—Generate test traffic, orders, account top-ups, and so on, and then compare OSS/BSS records to the experience/order of the test "customer."
2. Passive consistency checks—Reconcile the meter point to some other OSS/BSS data relating to the same event, such as comparing the gateway to a local

service network node or reconciling OSS/BSS data to alternate data sources such as SS7 probes, partner network records, and so on.

3. Configuration testing—Involves logical analysis of configuration data versus all possible event scenarios.
4. Trending/analytical review—Crude but simple technique to implement and gives an indicator of gross errors. It is most useful as an indicator if everything is generally well or if there are issues following a change in the way systems work.

Detecting Calculation Errors

Despite the extraordinary emphasis placed on detection as being the dominant model for performing RA, it is relatively rare for RA teams to try to detect calculation errors. This is because of the effort required to do so and the high likelihood of false positives. The key techniques when trying to automate detection of calculation errors are as follows:

1. Trending/analytical review/proof in total—Take summary totals and compare them to expectations. Expectations can be based on historic averages or rules of thumb for calculation, or might involve comparing alternative ways of aggregating essentially the same data. This technique is crude but effective at spotting gross errors.
2. Recalculation—Compare the actual BSS calculation to an independent calculation on the basis of the same transaction records but different reference data. This technique is very good for finding even small errors, where the cumulative value of many small errors may still be significant. However, the problem with this technique is the effort expended on creating independent data sets.

Collection

Unlike the other "C"s, automating the RA of collection is about anticipation, not detection. The aim is to anticipate which customers present a credit risk. This is done by analyzing information from various stages and sources:

■ Information obtained when the customer is adopted
■ Information about customer behavior after they start using services and their payment history

The aim is to make the best use of the data to highlight warning signs intelligently. Then proactive steps can be taken to limit risks. This is realized by limiting what kinds of contract or service a customer can sign up for, or by implementing spend limits, or suspending services when the customer's level of credit passes a predetermined threshold.

Using automated rules to check and control customer credit ensures greater consistency than in a purely manual process. It also enables vigilance over staff as well as customers. Embedding the use of this kind of automation within wider processes and workflows is key to its being effective.

Linking Theory and Practice

An overview based on the 4 "C"s can sound very abstract compared to the specifics being offered by a particular vendor's product. However, there is value in being able to succinctly and abstractly define what a tool is meant to do. This is because insufficient or confused requirements analysis and project planning are the key causes of failure when implementing new tools. Here are some warning signs to look out for:

- A lack of clarity about the risks to be countered by the tool
- A lack of clarity about project priorities
- Thinking constrained by looking for a single system/supplier to automate all RA capability, instead of being open to solving multiple objectives with more than one kind of solution from more than one vendor
- Failure to identify all suitable options/vendors/products at the outset of the project
- Limited understanding of relative strengths and weaknesses of each option
- Being over reliant on an external consultant with ties to a specific vendor
- Being over reliant on an external consultant with limited experience
- Copying what has been implemented elsewhere, without considering if it represents the best fit to the needs of the business

Those are the warning signs. Here is a list of what commonly goes wrong in practice:

- A poor match between the selected automated tool's capabilities and the customer's real priorities
- A failure to identify opportunities to reduce costs by utilizing technology already owned by the communication provider
- Unrealistic assumptions about the availability, robustness, and quality of data feeds
- Gaps in knowledge of the relevant business rules
- Unrealistic timescales and objectives for implementation, testing, and post-implementation modifications
- Poor communication between the finance and technology divisions
- Inadequate user training for the tools that are eventually implemented

High-Level Pros and Cons Analysis

Before going into any further detail, any RA department should analyze options for any kind of automation by performing a very short and high-level analysis of the relative strengths and weaknesses. This kind of analysis should be executed before issuing a proposal to external vendors, as it is essential to be clear about what their offerings include and what the alternatives are before discussing detail. Once detailed discussions begin with a vendor, it is very hard to step back and perform a sanity check on whether a very different kind of offering is better suited to the challenges faced by the communication provider. In particular, if the communication provider fails to include a wide enough selection of products, they will likely never consider some potential alternatives that may prove cheaper or be a better fit to their specific needs.

To illustrate the idea of a high-level analysis of strengths and weaknesses, here are some generic analyses for popular types of RA automation.

Test Event Generators

- ✓ Monitoring of capture, conveyance, calculation, and possibly even collection.
- ✓ Can be implemented quickly with low levels of integration if the RA team is willing to sacrifice diagnostic power.
- ✓ Can address quality of service as well as revenue integrity.
- ✓ Invasive: can perform logic-driven and need-driven testing and be set up to immediately test new changes.
- ✓ Good for switch-to-bill goals.
- ✗ Poor for network-to-bill goals.
- ✗ Narrow and small test samples.
- ✗ Low volumes of tests.
- ✗ Bottlenecks on testing of conveyance, calculation, and collection because of the need to generate actual events.
- ■ Dummy event generation can meet similar downstream assurance goals, while avoiding bottlenecks and sample limits.

Network-to-Bill RAdb's

- ✓ Powerful assurance that the value of network assets is being realized.
- ✓ Focused checks relating to a key area of leakage, especially for providers offering corporate data services.
- ✓ Highlights flaws in processes for adopting new customers and ceasing old services.
- ✗ Mismatches still need to be investigated, as an error may either be with the network inventory or with the billing system.

✖ May generate a lot of false positives where record keeping is poor but no leakage is taking place.

✖ Reactive only; not good for anticipating future errors.

■ The source of network data is important, and it must be reliable.

■ There is a risk of relying on flawed network inventory data.

■ Network interrogation creates a higher integration burden.

■ Crude reconciliations are easier to implement than full service feature reconciliation.

Tools for Trending and Analytical Review

✔ Quicker, easier, and cheaper than most other forms of automation.

✔ 80/20 rule: in the absence of other controls, trending can be relied upon to find 80% of problems for 20% of the cost and effort.

✔ Can be adapted to provide assurance across a wide scope of challenges.

✔ Can be applied to any reliable streams of data.

✖ High risk of false positives.

✖ High dependency on the skills of analysts to correctly interpret the results.

✖ Can only be applied where there is a reliable stream of data.

✖ May have no baseline for setting expectations.

■ High coverage, low cost, and relatively low diagnostic power makes trending an ideal complementary/backup tool rather than a sole line of defense.

■ Trending can be used as a technique to highlight gaps in other controls and/or to identify the priorities when considering the need to improve other controls.

Recalculation

✔ Can check significantly higher volumes of data than would be possible manually, with much lower risk of omissions.

✔ Enforces discipline over communication and interpretation of tariffs: internal, public and in private contracts.

✔ Can implement monitoring with just one data feed.

✔ Powerful tool for change control.

✖ Provides assurance only over calculation integrity.

✖ Significant burden involved in maintaining an independent version of tariff reference data.

✖ Will typically only check a (large) sample, rather than all data, because of processing overheads.

■ Becomes more cost effective if also used to support revenue maximization and customer analysis.

■ Can be used to support what-if analysis, customer valuation, competitive analysis, and so on.

■ The capability comes embedded in some other tools—for example, test event generators.

Six Rules for Selecting Tools and Vendors

The mnemonic "SPIRIT" captures six important rules when considering a choice of tools and vendors (including in-house development):

Scope: Set the intended scope of the solution by considering the 4 "C"s and also which product/revenue/cost streams are to be addressed.

Prioritize: Formulate a project plan for implementation where the most important things need to be delivered first, rather than just having a long undifferentiated list of requirements.

Internal capability: Ask around the business and investigate existing tools and technology owned by the company, as it may be possible to adapt or build upon technology the communication provider already owns and uses for other purposes. This can avoid the need for a new RA system, or at least reduce the cost of deploying any new RA technology. Things to look for include equipment used to test quality of service, existing data feeds, data repositories, and control tags.

Review all possible solution types and providers: Consider specialist vendors, leverage of business intelligence (BI) capability, systems integrators able to offer multiple solutions, the possibility of enhancing controls within the OSS/BSS systems, and the use of in-house resource to develop and/or adapt automation for RA.

Identify preferred options and create a shortlist: Use a high-level analysis of strengths and weaknesses to frame decisions and select suitable solution types and providers for inclusion in the shortlist.

Talk to peers in other communication providers: Even a short informal conversation can provide vital intelligence on a potential supplier or suggest an alternative solution not previously considered. Best of all they may tell you about their mistakes—so you can learn from theirs instead of making your own.

Advice on Requests for Information (RFIs) and Request for Proposals (RFPs)

Finally, here are some short snippets of advice on RFIs and RFPs because the exercise of gathering information from external parties is often undermined by a lack of internal information or an unwillingness to make decisions.

Before considering what to look for in the vendor's response, first think about what information you need to collate and provide to get useful and reliable answers. Start with a clear statement on the current "as is" RA environment, and what is the desired "to be" target state once the implementation is complete. State, as precisely as possible, the intended scope, objectives, and outputs from the tool. Make reference to key aspects of IT policy that could influence the deployment, such as attitudes toward security of data and how it may be transmitted from place to place. Lastly, identify the internal data sources that will need to feed the RA tool; an external vendor cannot do this for you.

When reviewing a response, look for and ask for structured phases for the deployment of any tool. It will force you to be clear on what needs to be done and makes it easier to flag what are the higher priority deliverables that should be covered in the earliest phase possible (and not quietly dropped by an underperforming supplier). Treat the solution as a whole and look for a statement about solutions and not about products. Evidence of future-proofing is a good sign that the tool will continue to be useful despite the inevitable changes to the business being assured. Demand clarity regarding interfaces so there is no confusion about what data needs to be provided and the extent of work required in preparing the data feeds.

Finally, make sure there has been adequate two-way communication and validation about the expectations for data volumes and storage costs and also about the access, transmission, and timeliness of providing input data. Review the operational support requirements and ensure that the vendor can provide a level of support that matches expectations.

Conclusions

Many a RA department has underspecified or under-researched what they want or how to get the RA automation they are looking for. Executives are unforgiving if it turns out that the RA department has fluffed a costly project and has cost the business more than the benefits they deliver. Avoid the pitfalls through simple and complete planning. This planning need not be onerous; in fact, it often saves a lot of wasted time talking and reading about irrelevant detail. Use simple methods like the 4 "C"s, the high-level strengths and weaknesses analysis, and SPIRIT to structure your investigation of the options available. Techniques like this help to improve decision making and avoid the cost of false assumptions or miscommunications about expectations.

What Revenue Assurance Tools Get Wrong

David Stuart

I have worked on all types of revenue assurance (RA) tools from those procured from vendors to those built in-house by RA teams, and in my opinion tools fit into one of two following categories:

1. Those developed by software programmers (ITRA)—generally with little or no RA experience
2. Those developed by RA professionals (RARA)—generally with limited software development skills

Generally speaking, the ITRA systems are those sold by vendors and RARA systems are those built in-house (however, I have seen instances where this is not the case).

The two types of systems differ vastly in functionality and capability. First, if we look at the ITRA system, fundamentally, it is built to be reused; therefore, the system needs to be generic so that the cost of integration can be minimized. To do this the back end of the solution is usually built into the following three components:

1. Staging—This is the bespoke area of the solution where raw data is loaded
2. Core—This is the generic table structure into which the stage data is squeezed
3. Reporting—Rules are built (in some cases these are predefined, in others these are customized) to identify and report the anomalies (leakage) and a rich predefined GUI sits over the top providing a representation of revenue leakage

As this type of solution is developed by IT professionals it will be built following a common structured programming methodology that incorporates all of the capabilities of the database software they are using; that is, fields are all defined and predefined database functions are used or even built by the programmers.

The advantages of an ITRA system are that it is easily scalable and, in theory, can be amended with ease to fit your ever-changing business. Also, from a management perspective, it can give you a one-stop view of the deemed revenue leakage. They are, however, seen as quite expensive from both a capex and opex point of view. From the RA team's view, they do not provide enough flexibility to find all of the leakages.

Now, if we look at the RARA solution, we will see a system that is developed over time; it starts with one reconciliation and a couple of data feeds and slowly grows to become an all-encompassing RA solution. The back ends of these solutions are generally only a couple of the following components:

- Raw Data—As with the ITRA solution, this is where data is loaded.
- Reporting—As everything is custom-built, rules are run against the raw data tables. A GUI may exist, but is not deemed 100% necessary.

Please note that there is no need for a staging area for the data as there is no need for the solution to be integrated with multiple types of OSS/BSS.

As this solution is developed by RA professionals, table structures are usually all VARCHAR (as RA people know that no matter what the source of the data, it can be corrupted; therefore, VARCHAR is the only guarantee that it will get into the RA system). There are also a lot more rules against each data set, so rather than just being reports on set A records missing from set B and vice versa, the rules will also be looking at the integrity of each of the data sets; that is, show all records where the activation date is not a valid date, show all MSISDNs with less than x digits, and so on.

The advantages of RARA systems is that they generally do what the users want them to do; that is, provide all data at the most granular level, allowing the analyst to write infinite amounts of SQL queries to assess all aspects of the chosen revenue stream. However, from a management perspective, it is quite often seen as an unknown entity, as no one-screen dashboard generally exists. Because of the way that they are built they do become more and more difficult to maintain over time and a single major change to the OSS/BSS infrastructure can often result in significant downtime for the solution.

So when asking the question "what is wrong with RA tools," I generally find that the answer can be predicted on the basis of the respondent to the question and the type of solution that is in place, that is,

1. The RA analysts—love RARA and hate ITRA as their number one requirement is for the system to allow them to delve deeply into the data, while
2. The management team—love ITRA and hate RARA as their number one requirement is to have a snapshot of areas of issue and financial loss/recovery

There is of course the RA manager who, generally speaking, sits somewhere between the two.

Finally, in case anyone was wondering—what do I prefer? Well, personally speaking, I prefer any team I am working with to have gone through the experience of building their own solution. The reason being that by building your own solution there is a greater understanding of the complexities of the data involved and as a result the team's root cause analysis skills are better honed. It should always be remembered that no solution will do root cause analysis for you, a solution only identifies issues, and these issues should always be assumed to be data integrity issues until proven otherwise. Ultimately, though, these RARA solutions need to be laid to rest as the support and maintenance overhead increases exponentially over time; therefore, the procurement of an ITRA solution will free up resources to concentrate on the more complex aspects of RA, while still providing assurance over the core revenue streams.

Anyone Want a Revenue Assurance Tool?

Mike Willett

Fear not, this is not a classified advert, trying to sell a previously loved revenue assurance (RA) tool with only one former owner.

The question is: to what extent is it essential that the software tools that RA analysts use to identify leakage be designed with RA specifically in mind? In the early days of RA in an operator, I am sure many of us started with multipurpose desktop applications such as Excel, Access, ACL, and so forth and then, assuming the business case was proven, we were able to move up to a "proper" RA application, sitting on a nice big server and letting us grow the volume of data we could analyze, increase the speed of processing, build repeatable RA analytics, and provide coverage across more revenue streams, and so forth.

It is always interesting to observe how RA tools are often represented as solving known problems to find untapped revenue potential. It makes sense to state that "by using tool A, operator B was able to reconcile revenue systems C with D and found E millions of dollars." So by induction if you deploy tool A, then you can do the same type of reconciliation and find E millions yourself, or you might not find E millions but at least you can tell your management, with confidence, that C and D are functioning effectively.

However, knowing what has happened in the past does not always guarantee that we can know what will happen in the future. First, it seems that recounts of the past seem to be represented very linearly. By that I mean, A did B, C did D, but because E then did F, we had a revenue loss. In the present we seem to rarely move uniformly and seamlessly from one activity to the next but our understanding of the past, and what may or may not have been relevant seems to be predicated on diligent pursuit along a steady and unwavering path. Second, if looking at the past could predict the future, then the global economic crisis, for example, should not have been surprise to anyone. Simply put, the past does not always provide adequate indication of the future and we should see this in RA—B and C do not reconcile and we lost E; therefore F and G will not reconcile so we will lose H. These sums rarely add up as expected and if they do it is probably more by chance than good modeling.

If a revenue assurance tool has been designed for RA, then by definition, its functionality must be built on a combination of historical knowledge of loss as well as prediction of where leakage will occur in the future. What many service providers are going through now is a period of radical change across their products and services, the infrastructure that supports them and the processes that are wrapped around them. The uniqueness of everyone's operating environment means that new and unexpected (and so by definition "not predicted") issues are arising more, rather than less, frequently.

My contention is that what is needed from an RA tool is not one that necessarily predicts where loss is, but one that is prepared for loss. By this, I mean one that has the necessary agility to provide timely and insightful analysis into an operational area and to drive greater insight into what is happening. This is about flexibility in the ability to read all manner of data, configure the appropriate logic, and provide a meaningful output. If we could predict where the future loss would be, we could address this before it happens, but our predictions are rarely accurate enough to enable correctly targeted, preventative action to be implemented. The tool that is required needs to find both the "we think something is wrong here" issues, as well as the "we did not even think that when we did action A that action B would occur."

To return to my original question—RA analysts do need a tool, but they need a tool predicated on the belief that assumptions of past loss will address only some of the issues. What is really needed is a tool also ready to handle the next generation of new and unpredicted business problems.

Automating Resolution

Eric Priezkalns

Lots of vendors sell lots of software, hardware, databases, and more that will automate the detection of revenue leakage. A lot fewer products are focused on automating their resolution. But if you never fix the leaks, then the value of finding them is zero.

The trend in recent years is for the sellers of detection tools to throw in some kind of software to manage resolution workflows as a freebie. This is great if you want to get all your tools from one supplier, but not so good if you use a variety of best of breed solutions. Adding resolution doubtless increases the value earned from the tools, and probably helps with both the initial sales pitch and building of customer loyalty afterward. One of the biggest downsides of integrating resolution into detection tools is that it encourages the revenue assurance (RA) team to behave as if the only way to detect leaks is to use automated tools. A little bit of imagination, the right selection of data, and some skills for querying and analyzing that data may be all that is needed to find a leak. Overdependence on vendors can encourage a team to feel that RA always begins with an expensive procurement exercise.

One of the reasons not to automate resolution is that it is only cost effective to automate what you intend to do over and over again. Automating the repetitive fixing of very particular faults implies that the root causes of those faults will never be addressed. You could say that in such circumstances the RA team is guilty of institutionalizing leakages. Unfortunately, the business case for fixing a root cause may be harder to make than the business case for addressing the symptoms and leaving the root cause in place. For resolution to be automated effectively, it has to be flexible enough to provide workflows that deal with all kind of faults. Those workflows have to be adapted quickly for every kind of leak, because you do not want to put a lot of time and effort into them. The more time and effort it takes, the harder it is to keep motivation on solving root causes, rather than just symptoms. In the end, RA has to be about solving all problems faster, not about finding ways to make them last.

Proof before You Buy

Eric Priezkalns

There are still many telcos who buy revenue assurance (RA) software without demonstrating that the software can be made to work (and deliver some value) first. No matter how rigorous a competitive tender process is, there is no substitute for a proof of concept. A proof of concept demonstrates that the vendor's wares work. Any reliable vendor will be happy to show that their product really can be made to work for the potential customer. Better still, get a proof of value that proves not only that their solution works but that there is some value to be gained by deploying it—though you cannot blame the vendor if there are no leakages to be recovered in your business. Although the vendor's software will be implemented for only a limited scenario and for a limited period of time, there is no better way to confirm the effectiveness of the RA product than to see it in action. It also shows that the vendor's staff has the necessary technical and management skills to realize success. Best of all, it gives the customer an insight into what the eventual financial benefits will be.

But a proof of concept is not just of benefit to the customer. A proof of concept also helps the vendor. It gives the vendor a heads up on how easy it will be to work with the customer's staff and to integrate systems in practice. Communication provider staff who are unhelpful, or who have limited knowledge, will probably lead to a more extended deployment project, pushing the vendor's costs up. Any proof of concept that utilizes real data will demonstrate whether the communication providers can be relied upon to supply the data needed to achieve their own goals. Good vendors have no reason to resist delivering a proof of concept before signing a deal; only poor vendors will want to rush to the sale and avoid a proof of concept.

Duplicate Data, Duplicate Costs

Eric Priezkalns

You cannot have a debate if you only hear one side of the story. Unfortunately, sometimes you only get to hear one side. Narrow-offering revenue assurance (RA) vendors argue they offer inherently better solutions than can be realized using business intelligence (BI). To say this conveniently ignores the fact that delivering RA by way of BI can significantly reduce the time and cost of realizing RA. I question whether purity about RA source data, the seeming justification for not using BI, is really better safeguarded with a stand-alone RA system. At times, the primary argument for a dedicated RA system seems not to be about the system, but to reinforce the need for a stand-alone RA department that will own, control, and use the stand-alone RA system.

Should RA data only come from primary sources? That is a wonderfully purist attitude, if you have an unlimited budget to spend. But it is not good business practice. The theory goes that only data from a primary source can be trusted. All secondary sources, so the argument continues, are less likely to be reliable. You cannot make business decisions on universal assumptions like this. Secondary sources may be just as reliable as primary sources. RA professionals often use secondary sources of data, and always have. Of course, secondary data may have been manipulated, processed, aggregated, filtered, and mediated in all sorts of ways. But then again, the same thing will have happened to so-called "primary" data. Manipulation of data is not intrinsically bad, and usually some degree of manipulation is a necessity for RA or any other task. The argument goes on that RA people cannot trust extracts of data if they did not design and control the original extract or the data repository. That is nonsense. Understand the data. If it meets the RA requirement, then use it. Save the company some money by avoiding the need to create duplicate extracts and duplicate repositories of duplicate data. If you do that, you avoid duplicate costs. There is no science that justifies making an *a priori* distinction between primary and secondary sources of data.

There are several reasons why most RA projects avoid using the same data as available elsewhere. The first reason is cultural. It is the "not done here" mentality. With this mentality, if something is not done by the RA team, it cannot be trusted. Excuse me for pointing out something that vendors and RA professionals often do not want to hear, but just because somebody comes from an RA background does not make them better at extracting and maintaining data than somebody who does not specialize in RA. People who work in RA make mistakes too. By the same logic, if somebody is an expert in extract-transform-load (ETL), but not RA, it does not mean they will make a mess of the ETL for an RA project. Sadly, there are many people in RA who are not experts in ETL, or in other aspects of large-scale data management, and some of the cultural arguments about data integrity seem to play upon the ignorance of RA staff. A second extract of the same data is just as likely to be faulty as the first, if you do not understand the cause of inadequacies with the first extract. If there is a genuine concern that secondary data is unreliable then RA people should help the business understand why its data is unreliable, and, if it is not fit for purpose, help the business to fix it. Taking the attitude of "we don't do that, we just do this" may help RA to keep a tight focus on priorities, but only at the risk of adding unnecessary additional costs to the business as a whole.

The second reason for duplication of data is control. RA people tend to fear that changes will be made to data, to software, and to hardware, without their knowledge. This will undermine, interrupt, or invalidate their work. This is a genuine concern backed by real-life examples, but duplication is only a short-term solution. Duplication of data means duplication of maintenance, which means duplication of effort. At some point or other, an efficient business will question whether the extra costs are justified. Either those costs will be overtly cut, by removing the duplication at some later stage, or the business will continue to waste money on maintaining the duplication, or, worst of all, all budgets are squeezed but no clear decision is made. These budget constraints may lead to errors and problems with either stream of data. Those errors could have been avoided if cost synergies had been realized and effort was focused in just one place. It is often less risky, in the long term, for businesses to maintain only one version of the same data (with backup of course), and for this data to be used by the widest group of users in the business. The business can get better returns and better integrity by focusing its investment on checking and maintaining the integrity of that one enterprise-wide version of data, than on distributing its money and effort across multiple versions, where each different version is used by a different silo in the business. In addition, wider usage means that more people, with varying knowledge and perspectives, are scrutinizing the data on a regular basis. This in turn increases the probability that any errors will be identified and remedied.

In the end, most RA people will undermine any pseudotechnical arguments about independence of data through their own actions. They do so by using the same data for two, three, four, or any number of different tasks they may have, so long as they are all under the control of the same RA department. This may be comforting if you think RA has a special place in the business, but it only serves to turn RA into a silo. Silos in business often end up looking like dead ends, as various experienced RA practitioners have found. In contrast, using the same data as other business users helps to ensure RA is integrated into the rest of the business. Using the same data for multiple purposes also makes good and basic business sense. Reuse reduces the cost of software, reduces processing overheads, reduces the need for hardware, and reduces maintenance costs.

The truth is, even the narrowest of niche RA software vendors will also use the same data feeds, the same data repositories, and so forth, for lots of different purposes, when it suits them. So do their rivals. Doing so is not a weakness; it is an advantage. Smart vendors are open about the advantage of using data for multiple purposes to serve the business. When I interviewed Benny Yehezkel, Executive Vice President of ECtel, for my podcast, he used the analogy of Airbus to explain ECtel's strategy toward data. As Benny explained it, the ECtel strategy is to carry data across the long haul to major hubs, and then shuttle data onward to the multiple endpoints. This makes good business sense. If ECtel, or any other vendor, had set up a new data repository to do fraud management, and you needed the very same data to also do RA, would you really insist on, and then pay for, the creation of separate, duplicate, data extracts and repositories? Of course not. In fact, if any vendor won a project then later pitched for a second project, they would always offer a cheaper price on the second project by proposing that the customer leverages the data obtained during the first project. Like elsewhere in business, a ruthless examination of commercial priorities will quickly brush aside any pseudoscientific dogma.

There is a third reason for why RA projects include duplication of data—and, hence, of costs: it helps vendors to make more money. They get paid for doing things, so generally they will recommend doing more, not less. If this means duplicating data extracts and repositories, then the money they get paid is worth just the same in their bank account. So, I am not surprised, but disappointed, to see vendors recommending waste, when they are supposedly selling business optimization.

If RA is to survive in the long run, it needs to be based on real efficiencies, not the wasteful exploitation of communication providers to make a quick buck. That is the other side of the story, and RA practitioners should hear it.

Chapter 7

Managing Risk

Fundamentals of Risk for Revenue Assurance Managers

Eric Priezkalns

A risk contains two components:

1. The impact, a bad thing that may happen
2. The probability of that bad thing happening

In English, the word is used ambiguously. It may sometimes be used to refer to both components, or to either component in isolation. Revenue assurance (RA) practitioners will benefit from careful and consistent use of the terminology to avoid confusion.

For RA, a risk is the mirror of a control objective. We can and should seek to identify a control objective for every risk, in a one-to-one relationship. Every bad thing that could happen can be addressed by implementing controls to counter its occurrence or the scale of the resulting events. The priorities will be determined by the seriousness of the bad thing that can happen, and by the probability of it happening. Of course, the bad thing that RA practitioners usually worry about is typically called "leakage."

Uncertainty

We do not know everything. Sometimes risks may be hard to estimate. Sometimes they may not be identified. When evaluating risk, we may need to be guided by

171

relevant expertise in the company. Sometimes we need advice from external parties. Our point of view can change as we learn more. It is very important to be open minded about the danger that a genuine risk has been overlooked.

We seek to get reasonable assurance that bad things are not happening and that processes are correct. We do not intend to get certainty of knowledge and should manage expectations accordingly.

Issues

The relationship between a risk and an issue is that a risk involves a probability; with an issue the negative outcome has occurred, is known, and is certain. Risks tend to be about the future, though risks may relate to events in the past or present if there is incomplete information and it is not known if the events have taken place.

A Simple Way of Prioritizing Risks

One way of ranking risks is to use a simple chart that compares probability to the severity of the impact.

We may use this table (Table 7.1) to sometimes help us prioritize. There will also be risks that do not fit neatly into this chart. For example, we may encounter problems that have a very high degree of probability and a very low degree of impact each time, but where the cumulative impact is hence significant. This extreme kind of high-probability, low-impact risk is very common in RA for communication providers because they deal with a high volume of low-value transactions. An example is an error in rounding of the duration of calls. In addition, we should be mindful of dangers that are very unlikely but where the consequences would be intolerable. For RA, an example might be an extended outage of the billing system coupled with a loss of transaction data. With these kinds of risks, RA connects to the goals of business continuity management. Blurred boundaries between disciplines like this tend to reinforce rather than undermine the assertion that RA is a risk management silo within a wider risk management universe.

We can extend the number of rows and columns in our chart, so there is more granularity in the probability and impact bands. Some charts use four or five bands

Table 7.1 Comparing Risk Impact and Probability

	High	Medium Priority	High Priority	Highest Priority
Risk Impact	**Medium**	Low Priority	Medium Priority	High Priority
	Low	Lowest Priority	Low Priority	Medium Priority
		Low	**Medium**	**High**
		Risk Probability		

for probability and impact, instead of the three shown here. However, there is no perfect way to group and sort risks. As with any technique, it is best to be mindful of its limitations.

Inherent Risk, Mitigation, and Residual Risk

In the absence of controls, the probability and impact of a bad thing happening is called the inherent risk. Controls reduce the likelihood of the bad thing happening, or lessen the impact when it does. This is called mitigation. The probability and impact after considering the effect of the mitigation is called the residual risk.

Our goal is to reduce risk, not to eliminate it altogether. Eliminating risk is impossible. Furthermore, assurance seeks to limit risk only to an optimal level. If the cost of mitigation is greater than the expected value of the benefit that results, then the mitigation should not be implemented.

Expected Value and Its Limitations

On the basis of the risk prioritization chart, it is tempting to extrapolate from its fundamental approach to a more exact method for prioritizing risks. If high probability and high impact means high priority, while low probability and low impact means low priority, why not simply rank priority based on the expected value of the risk? The expected value is the product of impact (quantified in consistent monetary terms) and probability. In a way, this technique would be more precise than using the chart in Table 7.1. However, using expected value has its weaknesses. The first is that there is no allowance made for uncertainty in the estimates themselves, so the precision of the expected value is spurious. The calculation of impact is likely to be an estimate. Calculating probability is likely to be worse than an estimate—it is likely to be an informed guesstimate—unless the problem recurs frequently enough that we can statistically analyze the history of occurrences. So while multiplying probability and impact gives a precise answer, it blinds us to the truth that expected value is derived from evaluations of probability and impact that are imprecise. This imprecision is inevitable while managing risk. Very few risks are so particular that there is only one imaginable magnitude of impact. When something goes wrong, the consequences may be greater or lesser, depending on blind luck. Giving a risk a single value for its impact and a single value for its probability is a shorthand for saying that we assume the impact and probability will be around the numbers we have estimated. We do not expect these figures to be spot-on. Furthermore, we can imagine the range of probability varying with the range of impact, so the same kind of risk is considered more probable when considering lower value impacts than higher value impacts. For example, consider the risk of a platform outage causing revenue leakage because transaction records are lost. The extent of the loss is linked to how long the platform is out of service. In turn, we can expect that a short outage would be more likely than a long outage, because we expect resources to be urgently deployed to get the platform back up and running.

Another potential obstacle to using expected value as a prioritization method is that not all risks have a quantifiable impact. This proposition is true of risks in general, but it is contentious whether it might be true of the risks dealt with by RA. If RA only deals with the kinds of risk where it is relatively easy to calculate the financial impact, then it would be possible to use expected value to set priorities for new controls. However, even where RA managers start out by thinking that their results should always be translatable into financial metrics, they often come up against counterexamples over time. One counterexample is when the scope of RA is extended to embrace delivering compliance to standards for accounting integrity. Here, the impact of a failure cannot be extrapolated from a simple measure of the scale of errors in the accounts. Another counterexample is where RA deals with both sides of charging integrity, and provides assurance that customers are neither overcharged nor undercharged. A simplistic use of expected value would either treat overcharging as a good thing (which would be wrong) or else it would say that all that matters is the magnitude of the error, so a $1.1 million undercharging error is ranked as being higher priority than an equally likely $1 million overcharging error. The problem here is that the scale of the error when overcharging does not relate to the ultimate impact on the business, which is damage to reputation, customer churn, cost of handling customer complaints, and so on. Some of these, like reputation damage, may be immeasurable. As a consequence, even a highly data-oriented discipline like RA needs to have a prioritization method that embraces qualitative as well as quantitative judgments about risk. It may be that even a small error could have a profound impact on reputation, and its treatment should be prioritized over a significant cause of leakage.

Revenue Assurance and Fraud Management within the Context of Enterprise Risk Management

Eric Priezkalns

Managing risk is not new. Many aspects of business involve managing risk. Managing risk just means looking into the future and taking cost-effective steps to reduce uncertainty where possible, and to guard against its negative consequences otherwise. Risk and reward is the fundamental equation of every business. Investors risk their capital in the hopes of generating a positive financial return. It is up to the company's stewards, its board of directors, to determine the right risk appetite on behalf of the investors, and to ensure that the company's actual level of risk is in line with that appetite.

As there are many kinds of risk—financial, strategic, reputational, regulatory, and operational, to name an arbitrary few; the job of measuring the total risk and comparing it to the appetite is not that straightforward. Communication providers are used to the phenomenon of finding and addressing new kinds of business risk, many of which are prompted by the complexities of their fast-changing industry. Fraud management and revenue assurance (RA) are two risk silos that were born of the particular risks faced by communication providers, and of the negative impact of not dealing with real risks of criminal exploitation and undiscovered errors in processing.

Large corporate failures mean there is increased pressure for boards of public companies to have a holistic view of risk for the enterprise and to understand connections between risks, so these can be managed on behalf of, and explained to, shareholders. The practice of enterprise risk management (ERM) has evolved as a solution that provides an overview of the varied risks that face the business; see Figure 7.1 for a sense of how it brings together disparate strands of risks to present consolidated information to the company's board of directors.

ERM is a complement to, not a replacement of, specialist risk silos. Both fraud management and RA have emerged as specialized and distinct disciplines precisely because skills and functions like internal audit and accounts receivable are underpowered to cope with some particular but significant risks faced by the business. The classification of fraud as a risk is uncontroversial. Revenue leakage is a risk too, albeit one with an atypical skew toward impacts that are low in value when seen in isolation, but attain a significant value when viewed as the cumulative consequence of common root causes. The advantages of specialist silos for key risks are as follows:

- Deep knowledge of often impenetrably opaque topics
- Tight focus on goals that less well-defined functions may lose sight of
- Specialist recruitment and development of staff
- Specialist automation

In the ultimate evaluation, if the risk is great enough, the cost of specialized mitigation is justified on the basis of the value it will protect for the business.

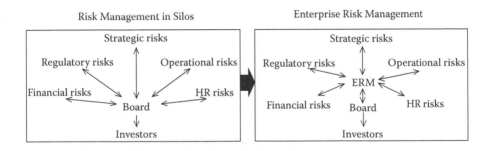

Figure 7.1 Silo risk management versus ERM.

However, while the benefits of silo risk management are clear, there are drawbacks to managing risk only in independent, isolated silos. The most significant of these is that the silos inevitably compete for the time, attention, and support of executives. It is often said that C-level sponsorship is vital for the success of RA or fraud management. While this is true, it is also true of many other activities, especially those relating to risk. The danger with having multiple uncoordinated risk silos is that ultimately they have some commonalities in the business case for what they do, but by being independent they counterproductively fight for resources. Without common communication channels, resources are sought without presenting a fair and systematic sense of priorities. Arguments for the value delivered by risk mitigation may be inconsistent, further complicating decision-making.

Another weakness of disconnected risk silos is that each silo can become a career straitjacket for the people working within it. The silos provide limited opportunities for advancement, and even fewer for sideways moves to other jobs that are still within the risk domain. A rigid sense of scope of objectives is especially limiting when it comes to making the best use of transferable skills. These career limitations are a factor in the increasingly frequent expression of a desire to evolve RA into "business assurance," although the phrase itself is so vaguely defined as to suggest it only means something more than the scope of RA without being clear about what is part of that "more."

Linking risk silos under the auspices of ERM has several advantages in addition to improving communication, prioritization, and the range of options for career development. Most crucially, it enables objective analysis of the gaps between the risk silos, where risks have been left untreated. This is another subtext of the growth of RA into business assurance—that there were also risks not being dealt with but that the RA department previously lacked the resources, will, or support to do anything about them. Just as there may be gaps in risk management, so there may also be overlaps. An effective ERM function will address these to avoid duplication and wastage. Turf wars should also be avoided, which can sometimes be difficult if specific risks become newly important to executives or are suddenly realized to represent a lucrative source of increased value if mitigated correctly. Collective action and common prioritization helps to avoid competition between risk silos. In particular, a consistent mechanism for evaluating risks and benefits of mitigation helps promote good decision-making and assists executives and ultimately the board of directors to correctly align risk to the corporate risk appetite. When considering the potential for biased reporting of risk, it is worth bearing in mind that bias can lead to either the underestimation or overestimation of both the risk and the benefits delivered by specific mitigations.

Guidance on how to deliver ERM within communication providers is very limited. An understanding of the risks faced by communication providers is as likely to be built from the floor up—with an understanding of specific risk silos like security, business continuity management, fraud management, and RA—as by applying generic ERM standards like ISO 31000. The TM Forum's eTOM places both RA and fraud management alongside other risk silos within their second-level grouping of ERM. However, in practice, we can see that this particular ERM construct is little other than a loose gathering of abstractly related activities that currently take place within most communication providers but that exhibit no signs of a common decision-making process or integrated framework. In short, they are put together in the eTOM for want of an appropriate place to put them, and not because the synergies have been identified in practice. This is most obvious when considering that some risk management activities are given an alternative home in the eTOM hierarchy. For example, management of regulatory risk is an aspect of regulatory liaison, and the eTOM places this within the context of managing external expectations. Managing customer credit risk is grouped with management of billing and customer relations. In part, this reflects a weakness inherent to the eTOM projection, where there is a strict parent–child hierarchy in the process groupings. Though this fits well with how a typical organization chart looks, it denies the reasons why businesses need the interleaving relationships of a matrix style of management, especially when handling risks that cut across organizational silos. In contrast, the intertwined responsibilities of both performing operational duties and participating in risk management suggests organizational connections that cut across traditional top-down organization charts.

The same factors that drive RA to move toward business assurance are driving communication providers to be more efficient and holistic in handling all kinds of risk. The RA and fraud management departments want to ensure that their risks and results have a reporting line to the top. Faced with the choice of trying to separate lines of communication or having a guaranteed channel as part of ERM, the advantages of pooling results into common reporting are clear. Part of the purpose of ERM is to ensure consistency in information; the valuation and probability of leakages will increasingly be determined across silos, not by silos, as analogs to revenue leakage are increasingly identified in other elements of the business. As traditional revenue leakages are plugged and the number and value of issues decline, the business case shifts from the certainty of getting value from historical issues to the uncertainty of preventing future issues. In other words, there is a shift in emphasis from reactive issue management to proactive risk management.

Though it is possible to make the case for "joined up" assurance that covers all enterprise risks, the model does not yet exist. The absence of a unified model that combines management of universal business risks with those that have very specific relevance to communication providers—like fraud management and RA—is not solely the fault of the TM Forum. No industry organization has yet brought together enterprise risks into a single risk management framework for communication providers, although it is to the credit of some communication providers that they have made great strides in doing this for themselves. Working in isolation is always problematic, as it stymies opportunity to learn from what others do and blocks the creation of a shared sense of best practice. In turn, the lack of the overall framework can encourage the phrase "risk management" to be overused. It can be applied to tools and techniques relevant to handling a narrow subset of risks. This may complicate explanation of how the communication provider needs a broad range of risk management tools and techniques to address the full array of risks.

Unified enterprise risk management would need to synthesize its approach by drawing upon multiple existing disciplines. In turn, this means working with disparate people from disparate backgrounds with disparate skills, but with a shared sense of purpose. Driving joined-up thinking in communication providers is always challenging. The obstacles are exacerbated by the turnover of technology, rapid growth, limited pools of experienced risk management practitioners available for recruitment, and low overall maturity of ERM in most communication providers. In some sense, ERM is currently in a similar situation to where RA was at the start of the century. Selling the benefits of ERM to skeptical executives is difficult, partly because the benefits are so intangible and difficult to measure, and partly because it is not possible to rely on a sense that they should do it because everyone else is doing it. Engaging consultants may not help; they can easily undermine the goals of ERM by selfishly pitching for small pieces of risk management in preference to making the big argument for holistic risk management. If communication providers are to develop the practice of ERM, they need to do so with the same pioneer spirit that has characterized the development of RA and fraud management.

RA and fraud managers need to be in the vanguard for realizing ERM within communication providers. Without their engagement, it will not be possible to manage the full spread of risks. On the other hand, they are naturally suited to taking a lead with developing the joined-up assurance model. They have succeeded in the challenge to establish a new discipline before. This gives them vital experience in the soft skills needed to influence, secure buy-in, and develop a team ethos that overcomes the barriers to change. Embracing a wider view of risk gives meaning and context to the otherwise nebulous idea of "business assurance." Crucially, if RA and fraud management functions do not take the lead, then the evolutionary path of ERM is likely to be top-down, as a consequence of the imposition of corporate governance standards and the demands of credit rating agencies, allied to the spread of worldwide ERM standards like ISO 31000. Waiting for a nonspecific risk management model to be applied to communication provider–specific risk silos might not be the wisest approach to securing the foundations of fraud management and RA. There would be potential for a culture clash between those with intimate knowledge of communication providers and those applying a generic understanding of risk. Of equal importance is the realization that risk management in communication providers requires very high end data skills compared to most other businesses, and much of the progress of RA and fraud management has been in deepening the understanding of how important this is to safeguarding the company's interests. Efficient use and reuse of data is a key common factor for success in successful risk management within communication providers. This places an emphasis on building a team with the skills to analyze data and the technology to obtain and explore data. Realizing synergies in risk management gives a route for staff with that extraordinary hybrid of skills and knowledge to increase their remit, productivity, and responsibility without the business needing to find a commensurate increase in budget that would come from recruiting people into more silos or by failing to realize a joined-up approach to assurance. For these reasons, communication providers should ask fraud and RA managers to take a lead in developing the joined-up model of managing enterprise risk.

Linking Revenue Assurance, Risk, and Maturity

Eric Priezkalns

If you ever go to a conference with a title involving audit, risk, and controls, and listen to a speaker from the revenue assurance (RA) community, you may enjoy what they say, but you may not notice what they do not say. The average RA speaker usually fails to talk about audit, or risk, and often fails to talk about controls. Most people who do RA see it as a special and stand-alone discipline, while good risk management is about avoiding a silo mentality. What gets described as best practice for RA often conflicts with best practice for risk management. This can easily happen in communication providers where you find people working in each silo with no overall boss who forces them to integrate what they do. Fortunately, I am lucky enough to be able to talk first-hand about one example of RA best practice that strongly advocates that RA evolves toward becoming a component of an integrated and enterprise-wide approach to risk management.

The RA maturity assessment, so long in development, identifies silo-based RA as an intermediary stage in its development. To reach the highest levels of maturity, the activity of RA has to grow beyond the confines of a silo and be fully integrated into the business. In turn, the nature of RA changes to become an element of enterprise-wide risk management. I know this will not be a popular or welcome message for some, especially those interested in building empires or fighting turf wars. But it needs to be said. If integrated risk management is not the destiny of RA, it can only be because businesses fail to take a holistic approach to risk. The operational risks within the scope of RA are not a special case. They need to be assessed and measured alongside all the risks the business faces.

Chapter 8

Understanding Controls

Control Frameworks and Their Purpose

Eric Priezkalns

We can primarily use the phrase "controls framework" to refer to the actual body of controls that apply to a specific domain, and also to the documentation identifying those controls and the need for additional controls and control improvements. Control frameworks provide a systematic approach to documenting, implementing, and improving business controls. A business control is a process that safeguards resources. The concept of a control equally well applies to revenue assurance (RA). The RA function not only implements, executes, and oversees controls, but can also manage its work within the context of control frameworks.

Information has a double value to the business. Operational information, such as the records of the individual services sold to customers, needs to be safeguarded to avoid mistakes and waste on a day-to-day basis. Mistakes with operational information will ultimately impact other resources such as the cash received from customers. The summary of information, such as that presented in management and financial accounts, needs to be correct so management and external stakeholders can make informed decisions. The impact of poor information on decision making is harder to quantify, but the potential downside to the business of reaching an incorrect decision will typically be greater than the downside of errors in operating data.

In an electronic communications business, a lot of effort is expended on the complicated job of managing information. Even so, mistakes are inevitable. For example, operational data may be corrupted or lost. A more insidious problem is that management data is created and is available, but nobody is sure who is supposed to act on it. A third kind of problem may involve the incorrect summary or interpretation of operational data for management purposes. Because information is intangible, and it is not obvious how to distinguish genuine information from misinformation, particular care is needed to design and implement effective controls around information.

Complex and information-intensive businesses such as communication providers use control frameworks to improve their business controls methodically. They do this by documenting the existing controls, determining their effectiveness, matching this to control objectives, identifying gaps, promoting new and improved controls as required, and testing that the controls are working in practice.

From experience, it becomes apparent that there are always differences in expectations between different audiences. That is normal, though it complicates giving advice. Being dogmatic may deliver a good approach for one business, but a terrible one for another. Even within a single business, there are inconsistent expectations. People may believe they have a common vision, only to find that, at some critical juncture, their different reactions highlight a subtle but fundamental difference. The understanding of controls in a business will be determined as much by actual practice as any words that can be communicated. Learning will come from practical experience. Being conscious of this does not mean we cannot adopt a fixed approach at a given point in time. It does mean that we should be open minded about what kind of approach is best suited to the maturity of the organization's control environment. We must consider the need to change that approach as the business becomes more experienced.

In the early stages of adopting an approach, it will be preferable to avoid an overinvestment of time in a theoretical framework. Some theory and principles will still be needed, but the early feedback from actual practice will likely have a strong influence on the precise methods adopted. This is an ordinary part of fitting a controls approach to the current business culture, especially as cultures will vary from business to business.

It is very likely that the controls framework approach will be updated following, and possibly during, the completion of the first few work cycles. Change should be scheduled on a periodic basis, rather than taking place gradually and continuously. The approach should be reviewed when it is apparent that major changes are needed and at least annually.

The likely evolution of the controls method will be to start with a limited range of controls objectives. A sensible place to begin would be setting objectives relating to the completeness, accuracy, validity, and timeliness of various kinds of transaction processing, such as output to the bill or to the general ledger. When starting, risk evaluation can be crude because there will be a surfeit of known and existing issues to be dealt with. As coverage of the controls in the business nears completion, the remit will be extended to address a broader range of controls objectives. Documentation may become more detailed. A more sophisticated approach to measuring risk and setting control priorities will become appropriate. Additional controls will be asked for and improvements suggested, even for areas that previously had a satisfactory controls rating. Inconsistencies in the way the approach has been applied will be ironed out. Work will need to be integrated and dovetailed with other risk management activities that are managed by separate teams. The application of control frameworks may be extended beyond core business functions to those that are more peripheral. The range of control objectives may eventually be extended to counter the full spectrum of enterprise risks.

A key element of the desired evolution of control execution will relate to span and type of controls. Wide-span detective controls will be preferred in the early days to give coverage and to reduce the chance that issues go overlooked. Over time, it will be appropriate to shift focus to preventative controls and narrow-span controls, to stop mistakes before they happen and to identify the root cause of issues and take remedial action more quickly.

To apply a controls mentality successfully to RA, it is necessary to divide the business into subsets and look at controls for each subset in isolation. Breaking the business into manageable chunks is vital at early stages of maturity. These chunks may be termed "domains." Even as maturity progresses, work needs to be effectively split between RA practitioners, meaning that even while connections are identified there are still some (potentially arbitrary) boundaries between the domains of controls that are separately managed and reviewed. The task of improving controls, prioritization, and measuring progress can be aligned to each of the separate subdivisions of the business. By breaking down the business into control domains, we can make progress and provide feedback to executive management without committing to an enterprise-wide approach at too early a stage. It may be that, as work progresses, executive understanding and communication of their goals and priorities may be refined or articulated in new ways. This may lead to a change in method. At some point, when goals have been broadly established and are no longer likely to change, it may well be appropriate to link all control frameworks and adopt a more holistic approach. In the meantime, it is more appropriate to identify and analyze controls on a piecemeal basis. We will tend to talk of having a different controls framework for each separate domain of the business, although the methods and principles applied will be the same.

Control Objectives and Priorities

Eric Priezkalns

It is important to have a clear sense of the objectives to be realized by controls. Objectives achieved by controls are called *control objectives*.

It is natural that most people will ask for control objectives to be described in a way that closely relates them to the details of what they already know ("just tell me what you want me to do"). Revenue assurance (RA) practitioners need to exercise skill in converting abstractions about control objectives into requests at a sufficient level of practical detail, and in plain enough language that the staff can understand what is expected. This means RA practitioners have to speak to staff using terminology and language the staff are comfortable with. This means emulating the mode of expression used by staff, and not forcing a pedantic RA or controls language upon them. However, RA practitioners should be precise in the logic and terminology of controls when discussing controls amongst themselves, to reduce the prospect of error or fuzzy understanding.

A control objective expressed at a high level can be decomposed into a number of more detailed control objectives. Having detailed statements of control objectives makes it easier to map the correspondence to specific processes and to establish which control objectives have been satisfied by existing controls. It also makes it easier to identify gaps that need to be addressed. However, a control framework cannot be based on an arbitrary list of detailed control objectives. A simple but common mistake is to compile an arbitrary list of detailed control objectives until no more can be thought of, and then to assume the list is complete. Starting from higher level, more abstractly stated control objectives, and decomposing them into detail is the best way to safeguard against the error of omitting detailed control objectives. At the highest level, control objectives begin with a short and simple statement of the overall business-wide objectives. Otherwise, there is the danger that

- The detailed control objectives are incomplete, causing the business to fail to identify and address real and important risks and issues.
- Some detailed control objectives are irrelevant, causing a waste of resources.
- The detailed control objectives contain unnecessary duplications, or possibly even contradictions, causing a waste of resources and undermining prioritization.

Many business objectives may not relate to the purpose of RA. However, starting with high-level business objectives is a useful exercise in positioning how RA supports the business and also flags where RA goals may be in conflict with other business objectives.

Overall RA objectives are best derived from a mix of business goals, augmented by RA best practice, common sense, and past experience of typical mistakes that lead to revenue leakage. There are good reasons to supplement the business goals because these goals may take some things as assumed. For example, it is unusual for a business to explicitly state a goal not to make mistakes, but this is because not making mistakes is taken as being implicit in realizing every other goal.

Revenue assurance practitioners can derive detailed control objectives by applying more general objectives to the detail of how the business works. There will be many detailed control objectives. Limits on resources means that we should always have a clear sense of priorities. Although the ultimate goal is to realize all control objectives, a sense of priorities enables time and attention to be focused on the areas that would deliver the greatest immediate benefit to the business in terms of improved performance and risk reduction.

The possible overall control objectives for RA fall into two distinct categories. Some RA departments will seek to address both. Some only seek to address one of the two categories. It is important to have a clear sense of the distinct categories, as preconceptions about the purpose of controls can lead to misunderstandings. We also need to be clear when communicating exactly which categories of objectives have been realized and which have not, and balance time and priorities between the two. The two categories of objective are as follows:

■ Business Validation: In simple terms, the goal of business validation is to ensure that the business is doing the things it should be doing (and nothing else). These objectives relate to improving the underlying performance of the business by reducing and removing potential for error, bottlenecks, wastage, inefficiencies, and delays. Goals include maximizing revenues, minimizing costs, improving cash flows and margins in line with the overall business strategy. The focus is on changing operational activities wholly within the control of the business. Given the nature of most communication providers, these objectives will be realized through a broad spectrum of operational, IT, and financial controls. RA, cost management, and order pipeline management are subsets of business validation.

■ Reporting Verification: In simple terms, the goal of reporting verification is to ensure that the information reported by the business tells the truth, the whole truth, and nothing but the truth. These objectives concern the reliability of information reported internally and externally. Reliable information is vital for good decision making. Information may be corrupted as a result of accidental errors or deliberate bias. Reporting verification objectives will principally be realized through financial controls. These objectives may be framed or supported by externally imposed compliance requirements, such as section 404 of the Sarbanes–Oxley Act. The key audience for the financial statements is the company's shareholders. Verification should also be performed for reports produced solely for internal audiences, though because of the volume of management information that may be utilized by different teams within the business, the controls practitioner must set sensible limits on where to offer formal verification.

It is easy to blur and confuse the goals of business validation and reporting verification. Because so many activities involve data, it can be hard to distinguish when data is being used for operational reasons and when it is being used to make decisions. One way of distinguishing the two is to look at the purpose, endpoint, or audience for the data. If data is used as part of an operational process, such as the data stored in a record of a customer's order, then checking the reliability of the data (the correctness of the order as originally communicated by the customer) is a form of business validation. If the data is being delivered to someone so they can review performance or make a decision, then it needs to be subjected to reporting verification. Of course, part of the reason for reviewing performance will be to manage the business. Verified information is an important enabler for management to exercise control over the business. However, we must distinguish the goal of verifying the information that is an input to a control over the business from that of executing the business control itself. It is possible to have satisfied reporting verification goals, but failed business validation goals, if information is reliable but tells us that there are fundamental problems with the business. Similarly, it is possible to have satisfied business validation goals, but failed reporting verification goals, if there is a control in place to monitor business performance but the reliability of data input to that control is uncertain. The idea is easiest to explain in terms of a metaphor. Business validation is like controlling the fundamental health of the business. Like a metaphorical thermometer, we use information, in the form of reports and measures, to monitor the health of the business. We also need to calibrate the thermometer to ensure it gives a reliable reading. Calibrating the thermometer means verifying that reporting is reliable.

Comparison with COSO Definitions of Internal Controls

COSO (COSO 1992) identifies three kinds of objectives for internal controls. They are as follows:

1. Operational effectiveness and efficiency
2. Financial reporting reliability
3. Satisfying legal and regulatory obligations

Business validation is equivalent to establishing the effectiveness and efficiency of operations. Reporting verification encompasses reliability of financial reporting, but also includes reliability of key information produced solely for internal use, as well as this published publicly. It is less common, but still possible, that legal and regulatory compliance could also be included amongst the objectives to be realized through the RA function's control frameworks.

Business Validation Objectives

There are many ways in which effectiveness and efficiency of operations can be improved. It should not be the goal of RA to tell managers around the business how to do their jobs. Chances are that a RA practitioner will know less about how to do a specific job than the manager who is responsible for doing that job every day. The focus of RA efforts should hence be to establish controls that prevent or highlight failures where such failures may otherwise be difficult to identify.

Four "C"s: A Useful Abstraction for Understanding Control Objectives

Controls are applied in a context. Some simple abstractions about the context of what a business does will help RA use consistent rules of thumb when determining the kinds of controls needed and the degree of risk inherent to the processes. RA will focus on controls over the management and processing of transactions. The Four "C"s is a simple way to categorize the kinds of processing applied to transactions with customers, suppliers, and business partners.

- Capture: Recording of the details of a transaction by the business. Examples include the taking of a customer's order via a portal, a computer generating a data record of a billable event, the update of a purchase order to reflect that the goods have been received or procurement entering into a contract with a supplier. Capture is the input of new data that needs to be processed and managed by the business to operate, provide services, and manage finances. New transactions will be instigated with the capture of data.
- Conveyance: Transmission of data between systems and/or between people. Conveyance involves transfer of transaction data between parts of the business, enabling all functions to act upon it correctly. As a result of conveyance, the format and medium used to store data may change, but the meaningful content represented by that data is not. Examples of conveyance include processes where people e-mail information from one to another, batch update of one system with data from another system, and filing paper copies of contracts in archives.
- Calculation: Modification or augmentation of transaction data. This is most commonly the result of combining transaction data with a source of reference data and may involve the application of an algorithm. Unlike conveyance, calculation changes or adds to data. An example would be calculating the charges to be billed to a customer by taking transaction data stating what has been supplied and multiplying by charges as recorded in reference data.
- Collection: Settlement of the transaction. Strictly speaking, this may be collection of amounts due, receipt of prepayments, or write-off of debts. Collection has similarities with capture, in that there is recording of new data (that the transaction has been settled) and conveyance, in that there may be an exchange of information, as well as money, with an external party to confirm that transaction has been settled. However, controls are different as some of the activities may depend on the processes and actions of the customer and possibly even on other external parties.

A straightforward example of the Four "C"s starts with taking a customer's order (capture). The order is transmitted to the provisioning system (conveyance). The status is updated when provisioned (capture) and transmitted to billing (conveyance).

A charge is applied for the first period when the order goes live (calculation) and a bill is sent to the customer (conveyance). The sale is also included in customer ledgers and in totals in the general ledger (conveyance). The customer pays (collection) and ledgers are updated again to reflect that payment (conveyance).

The Four "C"s can be used to describe the steps in processing transaction data. The first Three "C"s of capture, conveyance, and calculation can also be used to describe processing of semipermanent reference data. For example, a tariff plan will typically be stored in a rating or billing system as reference data. This data will also go through the stages of capture, conveyance, and possibly even calculation (though this is less likely). For example, a new tariff is devised by marketing and is recorded on a predetermined paper or computerized template (capture). The template is transmitted to a database used to store information (conveyance). The entries in the database are transmitted, at the appropriate time, to the outlets for communication of prices with customers, such as Internet portals, paper brochures, and guidance for customer services staff, and so on (conveyance). The entries are also the basis for data entry of reference data to the billing platform (conveyance).

It is sensible to separate thinking about the controls over reference data from the controls over transaction data. Reference data is typically an input that will determine if a system processes transaction data correctly or not. For example, if the tariff data is captured incorrectly, then the calculation of customer charges will also be incorrect. One way to visualize this is to think of the management of transaction data as a linear series of steps. The management of reference data is an input to the processes for transaction data (especially for determining how data is calculated and conveyed), and this can be visualized as separate linear series that join the transaction series at right angles. We can then implement controls over the processing of the transaction itself, or over the management of reference data. For example, we can sample a bill and recalculate it, to see if there was an error.

The typical control objectives for business validation are that information about capture, conveyance, calculation, and collection is

- ■ Complete
- ■ Accurate
- ■ Valid
- ■ Relevant
- ■ Timely

Report Verification Objectives

Report verification objectives should be defined with reference to how information is used. Information that is reported externally, to shareholders via the annual financial statements, is subject to an external audit and a well-understood series of expectations about what shareholders need to see. The expectations for reports used internally will not be stated in anything like as much detail, but the same principle applies of considering the use that it will be put to. This links to the concept of materiality, which describes how great an error needs to be to have an impact on the person using the information provided. While described in terms of numbers, materiality is not absolutely quantifiable as it depends on the impact on the user. When key decisions may be influenced by a small variation, such as might cause a loan covenant to be broken or a nominal profit to be turned into a loss, then materiality will be low and extra care is needed.

Typically, report verification will look to ensure that information exhibits the following properties:

■ Completeness
■ Accuracy
■ Validity
■ Existence
■ Relevance
■ Timeliness

Understanding the Parameters and Choices for Business Assurance Controls

Eric Priezkalns

As they say, there are many ways to skin a cat. Control objectives may be satisfied by many different kinds and many different combinations of controls. The best choice will be determined by the specifics of the business, the objective, and the options available. This section outlines the guiding principles for how the revenue assurance (RA) function can promote effective and efficient controls.

Preventative and Detective Controls

All controls can be classified as either preventative or detective in nature. For the avoidance of doubt, controls that might be described as corrective will be categorized as detective on the basis that they implicitly detect an error before correcting it, as opposed to preventing the error from ever occurring in the first place.

Preventative controls realize an objective by preventing a failure from taking place. If a preventative control is effective in design and is correctly executed in real life, it should be impossible for the control objective not to be realized, because failures are stopped from taking place. Examples of preventative controls include the following:

■ Constraints on the format and content of data that can be manually or automatically input to a system
■ Management approval processes before making a purchase
■ System and process design reviews before the implementation of a change

Although preventative controls should prevent failures, the degree of reliability of these controls may vary greatly. For example, an automated constraint to prevent entry of data in the wrong format will have a zero failure rate unless there is a malfunction. In contrast, a human preimplementation design review should be assumed to be imperfect because of human error and the difficulty of anticipating all possible failure points in a design. Although the control is fit for purpose, we can treat this as an example of the control not being executed perfectly. The failure rate for such a human design review will be heavily influenced by the skill, experience, diligence, and time available to the people tasked to perform it. The likely failure rate of a preventative control should be considered when determining the need for complementary controls that address the same control objective.

Detective controls identify failures after they have occurred. Monitoring is commonly used as a synonym for the execution of regular ongoing detective controls. Because detective controls identify failures after the fact, it is often easy to design them to produce measures of performance also. Some detective controls are implemented as ongoing measures of performance where the need for further investigation is only triggered when the measure strays outside of a preestablished threshold. Examples of detective controls include the following:

- System data and accounting reconciliations
- Analytical review of monthly results
- Review of upheld customer complaints

Detective controls cannot realize a control objective on their own. For the control objective to be realized, remedial action should be taken for all failures identified by the detective control, although sometimes correction can take place automatically without human intervention. In some cases, although failures are identified, it may not be possible or desirable to remediate and address all the failures that take place. For example, if services are undercharged, it may not be possible or desirable to raise subsequent additional charges to the correct customer or customers. The possibility and ease of remediation should be considered when deciding the appropriateness of relying on a detective control and the need for additional complementary controls to meet the same control objective.

Like preventative controls, detective controls may be more or less reliable. A line-by-line reconciliation of transactions between two systems will have a zero failure rate if the reconciliation correctly matches like with like without any spurious matches (perhaps the records each contain a unique identification code to assist in this regard). An analytical review, however, will only identify gross variances and there may be legitimate causes of variances that mask the extent to which errors drive the differences being observed. A sample-based check will be subject to sampling error. Human error inhibits the reliability of all manual checks. The likely failure rate of a detective control should be considered when determining the need for complementary controls that address the same control objective.

Control Design

Controls should be designed to be

■ Effective
■ Efficient

The effectiveness of a control is determined by how well it meets a specific control objective. It is hence important to have a clear idea of the objective that is meant to be realized by any control. It is not desirable to have controls for the sake of controls; there must be a goal that is being realized as a result of having the control. Controls may be ineffective for one of the two following reasons:

1. The design of a control fits the intended purpose, but the control has not been correctly executed as designed.
2. The design of a control does not adequately support the intended purpose.

If the design of a control is ineffective, the control needs to be redesigned or an alternate control put in place.

One practical and simple approach to control improvement is to focus effort during the RA department's first review of controls on those controls that are clearly ineffective in design. Once these have been addressed, it may be appropriate to raise expectations in general for subsequent reviews. As a consequence the RA team will apply a higher standard of expectations to all controls, so some controls previously deemed to have effective designs are recategorized as ineffective even though the design itself has not changed. To assist explanation of the findings to the control owners and stakeholders, it is appropriate to informally communicate to controls owners that they have controls that passed the review but that are borderline in effectiveness and so a need for improvement may be formally identified in future reviews of the control domain.

It can be argued that an ineffective control is not actually a control. An ineffective control does not contribute to the satisfaction of business objectives. In fact, an ineffective control is merely an unnecessary cost to the business. An RA practitioner should seek to add value both by improving and introducing necessary new controls, and by saving costs through eliminating ineffective controls.

Sometimes it may be tempting to describe a control as partially effective at meeting a certain control objective. To take a simple example, if there is a control objective to ensure the completeness and validity of metering output, then a control that checks the validity of the output on a sample basis, but does not address the completeness of the aggregate output, may be considered to have partially met the control objective. Describing controls in terms of partial effectiveness can be confusing and sometimes leads to misunderstandings. It is recommended that this problem be avoided by rewriting control objectives to avoid conclusions of partial effectiveness. In the example, the control objective could have been rewritten as two separate control objectives: one for completeness of output and one for validity of output. The control over validity can hence be deemed fully effective at addressing validity but is irrelevant to meeting the completeness objective. If there were an additional control that simultaneously addressed both objectives, it would be sensible to separately assess the effectiveness of that control in relation to each of the two objectives.

There may be many different potential controls to address the same control objective. In practice, it may be desirable to implement more than one control to address the same objective, especially where there is the potential that failures may not be identified by one or other control. For example, when implementing a new system, a preimplementation design review may identify and eliminate many potential causes of error in a simple and cost-effective manner, but other causes of error may go unidentified. Hence it is appropriate to augment the preimplementation design review with postimplementation testing. One typical combination of controls involves a mix of manual and automated controls, to give the reliability typical of an automated control, and the intelligence to spot exceptions to the rule that may be missed due to the assumptions made when designing the automated control.

It is good practice to address control objectives through a combination of preventative and detective controls. Preventative controls are intended to ensure that the right results are always delivered the first time. Detective controls give a transparent measure and confirmation that everything is working as it should. Preventative controls give greater benefit to the business by virtue of the fact that they stop undesirable errors, thus eliminating the negative consequences completely. However, it is difficult to measure the successfulness of preventative controls. Detective controls deliver less benefit as they permit errors to take place and then there is a cost involved in dealing with them and minimizing the impact caused. However, the a posteriori nature of detective controls means they are better suited to giving transparent feedback on the successfulness of processing and operations, including whether there were weaknesses in the preventative controls. A blend of the two types of controls hence delivers maximal benefits while also giving reassurance that the controls are working as intended.

As well as considering effectiveness, RA should also appraise the efficiency of controls. Controls should be designed with regard to the cost to implement the control and the value and probability of adverse impacts addressed by the control objective. It would be disproportionate to implement a million-dollar software solution to test the completeness of billing for a product generating revenues counted in the hundreds of thousands of dollars. Similarly, if a product has a simple tariff and a straightforward proof in total has indicated no problems with the calculation of billing charges, then there is little justification to implement a more sophisticated and detailed control. Although RA's mission may be to deliver effective controls that address all objectives, this mission is unlikely to be successful in the long term unless consideration is also given to the efficiency of controls. For this reason, it is appropriate to recommend the design of new and modified controls with efficiency in mind. It may also be appropriate to suggest that redundant and inefficient controls be replaced.

Some control objectives are more cost effective to realize with preventative controls. For other objectives, detective controls may be most cost effective. Which is most cost effective is linked to factors like the inherent likelihood of error, the cost involved in dealing with errors, and the importance of timely processing with no delays. When reviewing the design of controls, consideration should be given to whether objectives could be more efficiently realized using a detective or preventative approach. When reviewing controls, there is a danger that the reviewer will orient their thinking toward the types of controls that already exist and forget to consider that very different kinds of controls may be more cost effective. This is most acute when scrutinizing an environment where all current controls are preventative or all current controls are detective. Consideration should be given to whether radically different controls may be cost effective, especially if they have strengths that complement the weaknesses of existing controls. When gauging the cost effectiveness of detective controls, the likelihood that failures cannot be remedied should also be factored into the calculation. A detective control may be effective in the sense that faults are identified, but if the faults cannot be remedied, the financial loss to the business related to the fault should also be calculated. If other controls would allow the faults to be remedied, or prevented, then they may be more cost effective overall.

Duplication of effort should in general be avoided. After appraising controls, it may be decided that some controls are redundant because the objectives are equally well realized by other existing controls. Alternatively, it may be appropriate to retain all the controls, but change the frequency with which they are performed to give an optimal distribution of cost and effort for the extent of risk mitigation that arises.

Detective Controls: Diagnostic Power versus Span

Span is the concept that a control addresses a greater or lesser breadth of processes. For example, recalculating the charges on a sample customer's bill can be used to identify failures in the processes for calculating charges, but will not provide evidence relating to whether the services itemized on the bill are consistent with services actually supplied to the customer. In contrast, setting up a test customer that uses a predetermined extent and value of services and reconciling this to the final value of the bill issued will test not just calculation of charges but also the processes to record services supplied and represent them on the itemization. A control that tests fewer processes is relatively narrow in span, and a control that tests more processes is relatively broad in span.

Diagnostic power is the concept that a control may be more or less precise at pinpointing the cause of failure when failures are identified. Diagnostic power is inversely correlated to span. Broad-span controls have lower diagnostic power because there are more processes that could be the root cause of any problem identified. Narrow-span controls have higher diagnostic power because they address fewer processes and so there is a smaller potential locus for the root cause.

A simple way to understand the relationship between diagnostic power and span is to think of data that is transferred, and can be reconciled, between multiple systems. If there are 10 systems in sequence, it would be possible to reconcile the data between the first system and the second system, the second system and third system, and so forth. Each of these reconciliations (nine reconciliations in total) has a low span but high diagnostic power. Alternatively, it would also be possible to reconcile the data between the first system and the tenth system, without examining the data in any intermediate systems. This reconciliation between the first and last system would be broad in span, because it covers all processes at once, but weak in diagnostic power, because if a discrepancy is identified, the fault could have occurred anywhere in the sequence.

Detective Controls: Applying the Four "C"s

Table 8.1 provides simple and definitive rules for what kind of detective control is appropriate for each kind of processing activity as categorized using the four "C"s. By first categorizing all processing using the four "C"s, and then applying the corresponding detective control, the RA practitioner can build a consistent and robust framework of detective controls across all processing.

Table 8.1   Detective Controls Based on the Four "C"s Breakdown

Process Type	Detective Control
Capture	Comparison to an independent source of data about the activity being captured. Examples: • Mystery shopping • Multiple independent capture of the same event • Reconciliation with an external party's records
Conveyance	Reconciliation between the start and end points for conveyance
Calculation	Independent recalculation. This includes scrutiny of a sample of calculations by a person and automated recalculation using independent transaction and/or reference data.
Collection	Reconciliation of account balances is a must, but controlling collection is predominantly a matter of preventative controls.

Complementing the Span of Controls

Broad-span controls, while weak diagnostically, give assurance that nothing has been missed within their span. Narrow-span controls can sometimes lead to false confidence, even if many in number, if there are gaps between the controls that have been implemented. Errors may take place within those gaps in control coverage. However, narrow-span controls are less likely, in general, to deliver false positives because their logic is more specific to the processing being controlled. Narrow-span controls, because of their higher diagnostic ability, are more likely to lead to lasting resolution of root causes of problems, and place less burden on staff to investigate and follow-up issues when found. In particular, the follow up to problems identified by broad-span controls may itself make mistakes and may lead to misdiagnosis of root causes. Furthermore, follow-up investigation may be time consuming, meaning that urgent issues may not be resolved as quickly as they need to be. The relative strengths and weaknesses of wide-span and narrow-span controls may be balanced through using a strategy that incorporates both. A string of narrow-span and timely controls can serve to give pinpoint assurance over the processing chain. However, to ensure that all controls are working and there are no gaps, it is also worth, occasionally, executing a very broad-span "helicopter" detective control, reviewing activities over an extended duration to see if it can identify any areas where controls need to be improved. Even if no potential improvements are identified, the broad-span control gives confidence that the other controls in place continue to be effective and adequate.

When it comes to narrow-span and broad-span controls, the metaphor of a tightrope walker is appropriate. The tightrope walker's first guarantee of safety is the way he or she walks the rope. Small, careful steps reduce the likelihood of falling. But should the tightrope walker fall, he or she will want to fall into a wide net. In the same way, narrow-span controls are the first line of defense, but in case they fail, it is useful to have a catch-all broad-span control in place.

Conclusions

When dealing with detail, RA practitioners may sometimes fail to step back and consider the bigger picture. Controls should be understood in the wider context, with a view to ensure they deliver maximal benefits to the business. When seeking to improve the control environment, consideration should always be given to the following:

- Designing controls that are both effective and efficient
- Choosing to implement preventative controls, detective controls, or a blend of the two
- The span of controls and how this compares to the span of other controls and the extent of the domain to be controlled

Chapter 9

Serving the Retail Customer

Billing Accuracy Complaints and the Value–Visibility Curve

Eric Priezkalns

I have commonly heard that improving billing accuracy leads to reduced customer complaints. Although I have often heard this, and superficially, the connection seems obvious; the truth is that I have never seen any data, from a real communication provider, that supports the assertion. On the contrary, I have seen providers who greatly reduced the extent of error in their retail billing and account management, but seen the number of relevant complaints unchanged. Further analysis of publicly available data, such as that provided by the CPI scheme in the United Kingdom, further undermines any assertion of a simplistic connection between bill accuracy and complaints. I want to explore the reasons why in this article.

From personal experience, the following factors have explained why, at least for communication providers I am familiar with, the number of complaints are not well correlated to the accuracy of bills:

- Only a minority of bill accuracy complaints are considered justified.
- The majority of bill accuracy complaints are resolved, to the customer's satisfaction, by providing customers with more information without needing to adjust the bill.

- Some errors are invisible to the customer, such as small errors in measurement of duration.
- Some billing accuracy complaints are about issues with no monetary value, such as the accuracy of the customer's name or address details.
- Many customer queries are about understanding the tariff, but misapplying the tariff is only one kind of error affecting bill accuracy.
- Customer motivation is more complicated than we sometimes admit.

Value–Visibility Curve

The best way I could find to understand the kinds of complaints received by communication providers, and the issues that do not cause complaints, is by postulating that all customers exhibit their own personal value–visibility curve when it comes to their motivation to complain about problems with their bill or account. In Figure 9.1, the two axes are the value and the visibility of supposed issues—which includes problems with the bill that are both real and those imagined by customers. If the value is high, the customer will complain. Also, if the visibility of a problem is high, the customer will complain. The likelihood of complaint increases as you go higher on either axis, but a complaint can be prompted even if one axis is low so long as the other is sufficiently high. By visibility, I mean how obviously and easily the problem can be identified by a customer. For example, a separately presented one-off charge on an itemized bill is very visible to a customer. In contrast, a call that incurs the wrong charge because the start time was incorrectly recorded will be invisible to the customer unless the customer is in the unlikely situation to have independently measured and documented when the call took place.

The visibility–value model implies that the number or value of errors will not be well correlated to the number of complaints. It also implies that there will be no complaints about "invisible" errors no matter how large they are, and in general there will be fewer complaints about small errors even if visible to the customer. In particular, prepaid customers are unlikely to see information about charges in the

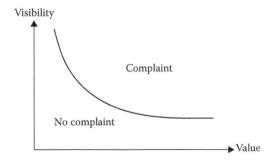

Figure 9.1   The value–visibility curve.

way that postpaid customers see the details presented on an itemized bill. This fits my observation that prepaid customers are far less likely to complain about charges than postpaid customers. In contrast, customers will complain about large charges, even when they are correct, because they are very visible.

The visibility–value model also allows for different customers having different complaint utility curves. From this we can observe that the numbers of complaints will relate to the average curve for customers, and this will vary from customer to customer and between different cultures around the world.

Though based solely on anecdotal experience, it is possible to perform a speculative plot of types of complaints (or lack of them) against a typical value–visibility curve. The items plotted on Figure 9.2 are as follows:

- The customer's name spelled wrong
- Monthly rental different to expectation
- Wrong rate per second applied
- Expensive (but accurately charged) calls—for example, roaming, premium rate services
- Actual calls missing from bill and itemization
- Small error in duration of a call
- Very large error in duration of a call
- Incorrect rounding

Lessons Learned

The priority for communication providers is to understand the actual drivers for actual complaints, instead of making assumptions. Perhaps the greatest single cause

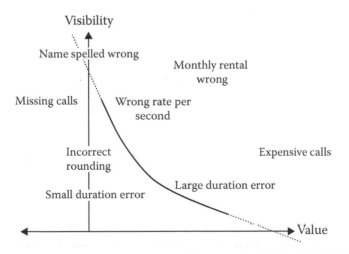

Figure 9.2   Plot of issues on the value–visibility graph.

of customer complaints about bill accuracy is that customers do not understand their tariffs. This means an easy way to reduce the number of calls to contact centers and complaints is by focusing attention on giving customers very clear and unambiguous explanations of how charges are worked out. It is vital not to jump to the conclusion that when a customer complains their complaint must have merit. At the same time, the provider should seek to identify their most valuable customers and appease them, irrespective of whether their complaint is justified or not. By the same token, identifying and distinguishing habitual complainants is a cost effective way to reduce costs. Instead of appeasing habitual complainants, it may be more profitable to save the cost of dealing with their complaints by effectively encouraging them to churn.

Using the Three "C"s to Tell When Customers Can and Cannot Protect Themselves

Eric Priezkalns

Figure 9.3 illustrates how electronic communications charging typically differs from many other industries and how this changes the nature of the relationship with the customer. In communications, if charging is not done in real time, a number of processing stages will create a cumulative delay between the capture of the sale and its display in the form of a bill or other account summary that the customer can check. This can be compared to a gasoline pump, a gas meter, or a grocer's scale, where the meter output is displayed directly in front of the customer. In many industries, the meter's output is exactly what is displayed to the customer. In communications, the "meter" output may be an automated data stream that

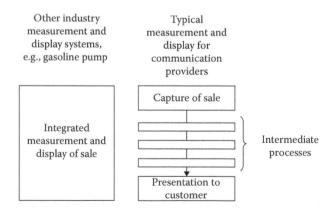

Figure 9.3 Comparing capture to display in communications and other industries.

needs further processing before it can be presented to the customer. This inhibits the customer's ability to check the accuracy of that output at the time the sale is recorded.

Unless communications charging is both executed in real time and presented to the customer immediately afterward, the customer has the extra burden of remembering what was purchased to verify that the charges they are presented with are both valid and accurate. Furthermore, the accuracy and validity of those charges are dependent on processes that occur after the sale took place. These processes are of significant complexity, meaning mistakes are possible. This is unlike other industries where post-sale processing of transaction data will typically be very simple or is just not needed because the customer pays at the time the goods or services are delivered. The difference is highlighted in Figure 9.4.

Because of the lag between capture and presentation, conveyance and calculation can be linked in the mind of the customer and in measurement. However, separately measuring conveyance and calculation is good practice. The causes of conveyance errors and calculation errors are likely to be distinct, and the two kinds of errors have a different impact on the customer. While a customer can always identify a calculation error if they receive an adequate itemization and are prepared to check their tariffs and the calculations performed, they are less likely to be able to identify a conveyance error unless they have a complete recollection of what services they used. Furthermore, they will not be able to distinguish a conveyance error from a capture error. This last point is important as conveyance errors may be corrected in a way that capture errors cannot.

Figure 9.5 illustrates, using a simple postpaid charging architecture, the distinction between monitoring for a conveyance error in isolation, and why monitoring of conveyance is often (unwittingly) combined with monitoring of calculation.

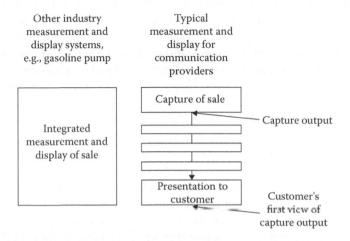

Figure 9.4   The customer's view of capture to display.

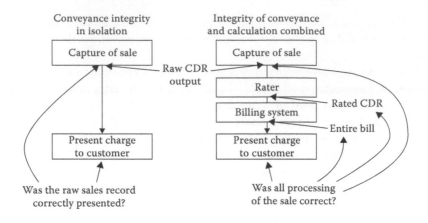

Figure 9.5 Conveyance alone and conveyance with calculation.

Conveyance error is measured by determining whether any corruption occurs to the data produced by the meter as it passes through intermediate systems. With calculation, the issue is that the algorithms for calculating the chargeable value are correct, and that the customer and tariff data that support these algorithms are correct. Separating the two concepts ensures that they are not confused and, hence, they are both properly measured.

One complication occurs when the itemization is not presented on the customer's bill. In this case, the communication provider's revenue assurance team can still perform the equivalent of a check of the bill itemization. Instead of checking actual itemizations, the data repository used to hold the meter data that would have been presented on an itemized bill and is reflected in the customer's account is checked. It is important that some repository of this type is installed and maintained properly because otherwise the business will be unable to respond to any queries regarding the charges it has presented and the balance on the customer's account. Such a repository is a precondition of analyzing conveyance separately from calculation. It should also be considered an obligation in its own right, as the business will be unable to justify its charges if it did not possess even this minimal form of record keeping.

Consumer protection issues are intimately related to what information is presented to the customer, and when it is presented. If the customers can "see" the capture output for themselves, they can verify it for themselves, then double-check any calculations, and hence assure themselves of all aspects that go to deliver the right charge. Small capture errors, such as might occur with an inaccurate grocer's scale, may not be detectable, but the customer will at least be able to observe gross capture errors. If the customer sees the capture output, but only at a later date, their task is made more complicated because they need to remember the actual events to compare this to the data presented to them. If the customer never sees the capture

Table 9.1 Time of Charge Presentation for Different Industries

	Time of Capture Presentation	Time of Charge Presentation
Gasoline pump	At time of sale	At time of sale
Electricity meter	At time of sale	Later
Itemized postpay communications bill	Later	Later
Communications bill without itemization	Never	At time of sale (prepay/real time) Later (typical postpay)

output then the onus is on the business to ensure fully that the charges are correct in every instance. In communications, customers who do not receive a full itemization are in effect unable to see the workings of the meters that capture and count the events for which they are charged.

Table 9.1 illustrates the different timings for informing customers about how much they purchased and how much it has cost them.

The reasons for not presenting customers with itemized breakdowns of their usage may be cost or convenience. But note that there is good reason to balance the saving in terms of cost with the increased obligation to protect customers from the consequences of errors that customers cannot identify for themselves. In addition, if customers cannot check the charges they receive, the communication provider loses a potentially useful source of feedback.

To summarize, the customers' ability to protect themselves, and hence their typical behavior when complaining about the integrity of charges, is related to the following:

■ The level of detailed information they receive confirming the sales for which they are being charged
■ The timeliness of presenting that information

Chapter 10

Advanced Themes in Revenue Assurance

Strength of an Argument

Güera Romo

As part of the literature review for my revenue assurance (RA) research dissertation, I contrasted the definition, objectives, and approach to RA, mainly using the work of the TMF and Mattison, supported by similar work by other authors. While most additional sources covered certain aspects of these dimensions of RA, the TMF and Mattison's work were more holistic in terms of the labels I chose to analyze RA from.

When I reviewed the material initially, I found Mattison's *The Telco Revenue Assurance Handbook* (Mattison 2005) and *The Revenue Assurance Standards* (Mattison 2009) very difficult to follow. The conclusion did not follow from the premises. However, these I can deal with in the analysis by comparing favorably those aspects that do add value and highlighting the inconsistencies or lack of cohesion where these are evident when summing up the contribution of a piece of work to this discipline. I included this work since it was one of the few sources that gave a holistic view of the entire function. It did not try to sell me tools or consulting services. Instead, it tried to explain WHAT the function is about and HOW it should be done. Exactly what I needed for this article.

I learned from the last review done by the university that the content of academic contribution is not always as important as the method used to arrive at such insight. The thought or argument journey we take to arrive at the facts is what differentiates facts from fiction, or in this case, from opinion. The methodology determines the validity and reliability of the contribution. We are all well versed in the art of RA and have experience, which makes certain things fact, like it or not. We will take on any academic journal and have it for breakfast because nothing makes up for the roasties we earned burning ourselves.

In an age of self-publication, why then do we still put so much importance on peer-reviewed publications and what exactly do we mean by peer reviewed? Surely when the Global RA Professional's Association says it ratified material amongst its members, this should constitute peer reviewed, as it is clearly a number of subject matter experts who reviewed and clarified the material before publishing it? The difference is in the rigor with which the review was done and the thoroughness of the methodology followed in writing the material as well as analyzing the output or result before publication. An objective and very critical analysis of content and indeed intent of the writing, by a third party, unemotional to the blood, sweat, and tears that created the initial output, culminates in a piece of work that can withstand any scrutiny. Such review will immediately eliminate an error of logic, an error of conclusion not following the premises. Peer-reviewed material also provides the opportunity to question the content and argue these until the facts hold up. This is difficult to attain when publishing a book or a non peer–reviewed article. However insightful such material may be, it does not provide a basis from which to build theory. It does not give one a bird's eye view on the different points of view, contrasted against and for an argument or the evaluation of support for and against the initial stances to such material.

Testing the Accuracy of Recorded Durations

Eric Priezkalns

The duration of an event is the difference in time between the time the event commences and the time the event terminates. Beyond this, there are many possible ways to define the duration of an event because of the choice of what is said to be the start and end of an event. For the purposes of determining charging integrity, there are two generic extremes:

- Input to the meter. The event commences when the answer signal has been acknowledged at the meter. The event terminates when the clear signal from the first customer to clear has been acknowledged at the meter or when the meter recognizes that the event is to be cleared for whatever other reason.
- User perception. This is the period of time that the actual service is available for use by the user.

There may be a difference between the recorded call duration and the actual time for which conversation is possible. Such a difference could arise, for example, due to delays between when a signal is sent and received. An estimate of the average transmission latency and its variability may be necessary to reconcile the user's perception and what is recorded by the meter. Certain types of delay may be introduced by deliberate design. Commonly these will be intended to ensure that a call is not terminated prematurely. It would not be legitimate business practice to implement delays for the purpose of increasing revenue.

It is reasonable to observe that it is highly unlikely that a typical communication provider would design its core networks to produce small but definite and unnecessary delays in signals. If the intended net effect is to increase the duration, the delays cannot be consistent. To systematically increase duration, the delay on the clear signal must be maximized relative to the delay of signals to establish the call. Prescriptive regulation and audit of signal delay should hence be considered unlikely to be of benefit to customers, and of no benefit to the communication provider. At most, a very occasional ad hoc practical test of signal delays would meet the same objective of protecting customers without the need for systematic and repeated compliance measurement.

Because of all the above, it is rational to choose a definition of duration that most closely represents the issues relating to the actual meter's accuracy. It also is appropriate to choose a definition that supports precise measurement using automated devices by minimizing the need to reconcile the duration per the meter and the duration per the measurement device.

Measuring Duration Error

Note that there may be a difference between each of the following three values:

1. The event duration as defined
2. The event duration as recorded by the meter
3. The event duration as measured by an independent test or measurement device

The purpose of measurement is to gain assurance over the accuracy of the event duration, though there is no empirical data that directly states what the duration is. The event duration can only be inferred from one or other of the two sources of actual data, that is, the meter or from an independent test or measurement device. The aim of a measurement strategy that addresses duration must be to compare the meter recording to an independent device. To avoid the impact of environmental factors, and errors due to the comparison device, the actual device should be designed to give the closest possible correspondence to the meter. This is necessary to maximize the assurance derived from practical, empirical observation of systems working in practice and minimize the reliance on theoretical explanations of the differences between the results.

There are a variety of different kinds of measurement or test devices that may be employed to similar ends. Rather than listing all the kinds of devices, it is more productive to contrast the main differences in design.

- Invasive and passive. Invasive devices create an event to check that meters work correctly. They record data about the event by virtue of creating it. Passive devices monitor real events, giving an independent output for comparison to that of the meter. They record data about events that they do not create.
- Near meter and near customer. Near-meter devices are located and designed to give similar output to the meter. Near-customer devices are located and designed to accurately record what the customer perceives. Devices that are near to the meter are to be preferred for measuring accuracy of duration.

Different devices may be regarded as "near" or "far" from the meter. These terms are used to reflect not just geographical proximity, but similarity of functioning. The following devices should be considered near to the switch and are the most common tools for testing duration of network events:

- Passive SS7 probes situated near the meter and used to monitor signals
- Invasive test generators that directly interface with meters

No measurement approach will give satisfactory results, if the devices used introduce errors because they are

- Flawed (the results they give are wrong even per the device's own specification)
- Imprecise (the results given are correct to a certain level, but not precise enough in the exact context they are used)
- Poorly aligned to measuring the exact property for which they are used (the results are correct and precise but the device's implementation and specification gives inherently different results to the properties being measured)

As such, sufficient care must be taken to address all three risks. In other words:

- The device must be tested to determine if it works as specified.
- The precision of the device must be determined both by checking the specification is adequate and by testing that the specification is satisfied in practice.
- The device must be chosen on the basis that it is well aligned to the properties being tested, or at least that any inherent difference is quantifiable and manageable.

Though appropriate steps should be taken to address these objectives, the nature of measuring duration lends itself to an additional, and generally more powerful, form of analysis on how well the measurement device is functioning. The basis of measurement of the meter is comparison to the independent device. As such, measurement is nothing more than repeated comparison of the output of the meter to the output of the measurement device. The measurement device is, somewhat arbitrarily, considered to give the definitive result for the purpose of measurement, and differences are wholly attributed to the meter. This is a sound approach for the purpose of measurement, as it assumes that all differences are due to the meter, so tends to report the worst case for the level of duration error. This is not quite true, because both the meter and the measurement device are producing an output that represents an ideal, for example, the actual time a signal is sent or received (as opposed to the time recorded by either device). There are hence two separate real variances between the ideal or "actual" property and each device:

1. There is a variance between the actual and the meter.
2. There is a variance between the actual and the measurement device.

As stated earlier, the cumulative difference is, for measurement purposes, assumed to be wholly attributable to the former variance, that is, to the difference between actual and meter. In other words, the latter variance is taken to always be nil. Of course, the real variance is a combination of the two and the latter variance may in reality be a nontrivial component. The two variances may be correlated or not. If correlated, then

- The specification of the operation of the meter is more similar to the specification of the measurement device than it is to the duration as defined; or
- The devices are not independent.

Independence should be readily demonstrable from the specifications of the devices. It is not in general likely that devices may be accidentally interdependent without that being understood as part of the implementation of one or other. So the likelier risk is that the specifications are more similar to each other than either is to the actual duration. For this reason, care is needed to define that property in a realistic way that can be used as the basis for assessing real performance. In effect, the similarity of the two specifications is a desirable quality, and if the duration measured is different to the two outputs, then the difference between outputs and duration is intrinsically environmental, as both devices are performing as intended. Clearly, when setting an objective it is pointless to set an ideal property (like duration) that cannot be precisely measured either by the meter or a test device. On the other hand, if quantifying the idealized duration is an aspect of protecting the customer, then the tolerances for performance must reflect any known environmental factors that would cause variances between meter output and the idealized duration. It should also allow for environmental factors between measurement device and idealized duration. This is because the only real comparison taking place is between the test device and meter. The idealized duration is not itself a datum in any real analysis, and so accuracy testing based on this property can only be hypothetical.

All of this discussion is, to a large extent, an overcomplication of the real problem. The theoretical problem is to monitor the variance between the metered duration and the actual duration by using a measurement device, while allowing for the difference between actual and measurement device. However, from the aforementioned it should be clear whether the meter and measurement device are independent. If independent, then the appropriate course of action is

- Compare the two outputs.
- If different, analyze the reason why as far as possible.
- If not different, take that as evidence that the meter is accurate.

Both the meter and measurement device may be different to the actual property in the same way and to the same extent. This is not a limitation on testing the meter, but rather a limitation on testing the idealized "actual" duration. The meter can be shown to work correctly per its specification by the measurement device without needing recourse to this idealized duration. If it cannot be shown to work correctly per the idealized duration, and no data can represent this point, then the problem is one of definition. In other words, the measurement can never state performance versus the expectations as defined because of the way they are defined. The practical solution to this problem is to alter the definition to fit practical reality, not to take practical steps to meet an unattainable ideal.

Use of "Forensic" in the Context of Revenue Assurance

Güera Romo

I have come across the use of the term "forensic" in the literature of the Global RA Professional's Association. I was unsure of its meaning and assumed it made reference to the analysis and resolution of fraudulent activity committed by either the subscriber or telco personnel. A cursory glance at the training material and the certification requirements also suggest a major focus on forensic skills. For example, in one course module the word "forensics" appears to be used as a synonym for analysis.

Gartner, in an article about setting up an RA Competency Centre (RACC), compared the approach of some eight vendor and consulting firms. In summary, all the respondents were aligned with regard to the need to understand the system and data architectures, as well as having clearly documented business processes end to end. All eight included a thorough analysis of the requirements in their first phase of the project. Some respondents were tool independent. No reference was made to forensics in either the Gartner scope of the project or the respondents' preferred approach and methodology.

My point is this. If the RA community wishes to standardize and unite this profession, we must speak the same language. I have not come across a workplace where this forensic approach to RA is present in the culture or language of the group. Individual enquiries as to its meaning returned either "the Global RA Professional's Association plays in the forensic/fraud field and not pure RA" and "the Global RA Professional's Association means RA as we understand it but it was written/coined by somebody who has a forensic investigation background."

The beauty of managing a team consisting of RA, fraud, and law enforcement personnel is that one gets a glimpse of their cognitive processing and respective approaches to problem solving using a single issue such as the perceived underbilling of wholesale data usage. There is a distinct difference in the mental processes of a forensic investigator and financial analyst. Both these personnel profiles can apply for an RA position. This is the sociocultural context within which we process our knowledge of RA standards and will necessarily lead to misinterpretation of an otherwise well-intended contribution. In change management we are taught to show sensitivity toward diversity. Is this question over the meaning of the word "forensic" a matter of cultural differences and diversity? If so, what can be done to establish a reliable translator for such term differences? If a job description asks for forensic skills when I really need somebody with an analytical ability, would I be doing a fair evaluation of suitable candidates?

IPDR-Based Usage Meter Accuracy

Eric Priezkalns

Cripes! Did I just write that heading? I surely did. When I use acronyms, they tend to be ROI, P/E ratio, BPR, or USP. In this article, for a change, I am going to focus on IPDRs and DOCSIS. Why? Because sometimes even I find that to do RA you need to roll up your sleeves, dive in the deep end, and start wading waist-deep through the technological soup of electronic communications. That is especially true when understanding metering accuracy, the dark art of RA that is understood by few and lied about by too many. Metering accuracy is the most extreme point on the network side of the network-billing continuum that RA people need to check for accuracy. RA people love love love reconciliations, but reconciliations are utterly useless if all you are doing is reconciling garbage in to garbage out. Checking for garbage in means checking the meter. Metering is the starting place for the most stereotypical RA discipline of all the RA disciplines—the assurance of the usage charges. But even this most ancient of RA checks has to move with the times. As the world and business changes, so more and more fixed-line communication providers and cable providers are considering making the shift toward usage-based charging of data services. That means reconciling usage per IPDRs, and, moreover, checking the accuracy of the data in those IPDRs. I want to explore the topic before any goons come along and make up some stats on usage accuracy, based on their own unrivalled nincompoopery. Trust me, they have done it before, and they will try to do it again. But I digress. Luckily for me, a report in the public domain means we can talk about IPDR usage accuracy, and even discuss real and tested stats, without revealing secrets and without relying on anyone's misinformation.

NetForecast, a network consultancy, were asked to audit the usage metering accuracy for Comcast subscribers served by the IPDR-based Cisco Cable Modem Termination System (CMTS) model 10000. The report was written by NetForecast boss Peter Sevcik and has been made public (Sevcik 2010). NetForecast's report discusses the factors that determine accuracy, and the reasons why the volume of data sent and received by the end user will differ from the volume of data sent over the network. For their audit, NetForecast performed a series of controlled tests, creating and measuring data traffic from the user's perspective, and comparing this to the metered usage per the IPDRs created by the CMTS. Their final conclusion was that the Cisco CMTS 10000 used by Comcast is accurate to plus or minus 0.6% over the course of a month's usage. National regulators should take note. It is no good setting accuracy expectations that are more stringent than this, because the end-to-end accuracy obtained by an operator will never be better than the accuracy of the metering equipment they use, and the equipment used by operators is manufactured and sold on an international basis by suppliers like Cisco. This places an effective limit on the accuracy attainable by any operator.

The Cisco CMTS 10000 reports usage in the form of IPDRs created every 15 minutes. The final mile of the user's connection is between the CMTS and the cable modem in their home. Differences between the usage recorded by the CMTS and at the user's home are hence down to differences between the volumes of data recorded at either end of the local coaxial or hybrid fiber-coaxial (HFC) used for the last mile. These can occur because protocols like TCP will cause packets to be retransmitted if lost in transit. If a downstream packet is lost between the CMTS and cable modem, it will be resent; meaning the volume of the missing packet is recorded at the CMTS but not at the user's end. If an upstream packet is lost between the cable modem and CMTS, it will also be resent; meaning the volume of the missing packet is recorded at the user's end, but not at the CMTS.

The NetForecast report also highlights a number of issues that may influence the user's perception of accuracy, as opposed to the actual accuracy. To begin with, when performing any kind of measurement, it is important to be precise about what is being measured. In this case, the volume of content that is of interest and can be measured by the user is only the payload carried within the traffic. There are also overheads relating to the protocols needed to carry that traffic. The DOCSIS specification defines how subscriber traffic is carried within Ethernet frames. Anything within those frames, be it the user's content or the overheads for protocols within the

Ethernet frame, will be measured by the CMTS. The essence of the NetForecast test was to FTP files of known sizes. NetForecast calculated that the FTP, TCP, and IP protocols added about 6.2% overhead to the traffic carried. So if a user replicated the tests by comparing the size of files they have FTPed to the volume of data recorded by IPDRs, they would see a variance of 6.2%. It is important to know that the overhead variance will occur and may need to be incorporated into customer-facing processes for customers who may complain about being overcharged for usage. Moreover, 6.2% is not a fixed amount for all traffic. The FTP protocol adds a low overhead, and many other protocols will add more overhead, leading to higher variances between usage as perceived by the customer and as recorded by the network.

Other factors that NetForecast rightly consider are timing differences because of how long it takes to produce, poll, and aggregate IPDR data, and the influence of rounding on the volumes measured. The average RA practitioner should routinely identify these factors for their business. The RA practitioner is more likely to overlook "background" traffic, which has nothing to do with how much the subscriber uses their service, but gets measured and added to the volumes of data all the same. In their tests for Comcast, NetForecast concluded that background traffic like SNMP polls and modem checks represented less than 1GB of traffic per month. In the context of Comcast's service, this was unimportant, as Comcast's usage monitoring is designed to identify use above 250GB per month, and there is a degree of inherent offset because monthly usage is always rounded down to a whole number of gigabytes. However, if background traffic was higher, or if pricing is more sensitive to use at lower volumes of usage, it would become increasingly important for the provider to manage customer expectations relating to charges that the user cannot influence.

Kudos to Peter Sevcik of NetForecast for writing a report that is clear, accessible, and covers all the important factors that determine the actual and perceived accuracy of IPDR-based usage metering of data services. Kudos also to Comcast for commissioning the report and for making it public. Transparency is an essential aspect of accurately and fairly charging customers. This report shows that while metering may be complicated, it does not have to be mystifying. Anyone else with an interest in accurate charges, and hence in accurate metering, should take note.

GSM Gateways: The Quiet Crime

Eric Priezkalns

Whenever people ask me about fraud in telecoms, I try to get one thing straight at the beginning. Do they want to know about the kinds of fraud committed by employees, committed by the "customer" (often in collusion with employees) or committed by other firms, some of which may wrongly be considered partners? In reality, fraud management departments within communication providers sometimes deal with only one or two out of the three, and leave the rest for somebody else to cover. It can make sense to split up responsibility for fraud. The skills needed to analyze call patterns for indicators of fraud are not very similar to the skills needed to monitor employees, which are dissimilar to the skills needed to understand the weaknesses that might be exploited by another communication provider. But splitting up responsibility, or focusing attention on one kind of fraud at the expense of others, can leave gaps in a communication provider's defenses. One of the biggest current gaps for wireless communication service providers (CSPs) is protection against the use of GSM gateways, also known as "simboxes." This gap in protecting against fraud often ends up being covered by the RA function.

GSM gateways are devices that allow a call on a fixed-line network plugged in one side to be connected to a mobile network on the other side. By bridging the world of fixed and mobile they offer a clever way to exploit price differentials of a mobile network provider. The fraud requires the use of a GSM gateway stuffed full of SIMs charged at standard retail rates. These get sited within range of a radio antenna and are used to connect calls to the victim network instead of using a normal fixed interconnection between networks. Instead of paying the full price to terminate an interconnect call legitimately, the offender instead pays the retail cost of a local call. This means the mobile network is cheated out of some of its revenues. In addition, concentrating traffic in one cell may lead to disruption of service for legitimate mobile customers. To counter poor service, the unwary mobile network operator may even find itself making an otherwise unnecessary investment in extra base station capacity. But this kind of fraud gets little attention. Why is that?

- The fraud can fall into a gray area legally. Contracts may not be worded tightly enough in stipulating that retail SIMs are not to be used by nonretail customers. In addition, legislators and regulators may not be keen to intervene. GSM gateways may lead to discounted services for the public, and are a back-handed way of eroding mobile termination charges without needing direct intervention in the market.
- CSPs using GSM gateways may be completely legitimate in most other respects. Few vendors or consultancies specializing in fraud and RA want to alienate potential customers, so often prefer to keep quiet rather than highlight this topic.
- GSM gateway fraud challenges most preconceptions about fraud and how to detect it. For example, there is no guaranteed link between this kind of fraud and bad debt. On the contrary, exploiters of GSM gateways may be mistaken for excellent customers, because they have very large bills but pay them promptly.

Like any kind of fraud, it is impossible to accurately estimate the impact of GSM gateway fraud. What we can say for certain is that vigilant mobile operators will suffer a lot less than those who do nothing to counter GSM gateway fraud.

A two-step approach is needed: tight wording of contracts to clarify that retail contracts are not available for businesses using GSM gateways as an alternate means of interconnection, coupled with constant monitoring and prompt disconnection of the SIMs in the simboxes.

Suffering a Crisis in Confidence?

Mark Yelland

I must be missing something. Is RA more safety critical than aerospace, or require higher reliability than a satellite? It must be, because both those industries have been using sampling strategies without problems for years and yet one never hears about sampling as an approach within RA. And I use these as examples because both have had high profile failures and yet neither has moved away from sampling as an approach.

So let us consider some of the benefits of a sampling approach for usage.

For usage-based products, using recognized sampling plans, such as the USA Military Standard 105E, for batches over 0.5 m, the sample size would be just 1250 to achieve an acceptable quality level of better than 0.01%, or 100 parts per million, which is probably good enough for RA. But that is less than 1% sampling, surely that cannot be right? According to well-established sampling theory, tried and tested over decades, it is. OK, you may not be comfortable that there is only a 95% confidence level that your sample will detect all errors above 0.01%, but you can always calculate the sample size required to deliver the confidence level necessary, but even at a 10,000 sample per batch level, it still represents a significant drop in volumes of data that need to be processed.

With less volume of data to be processed, the analysis speeds up and the visibility of issues is quicker, so the potential impact of any problem is certainly no more and potentially less than 100% sampling. With less time performing data analysis, the analysts can focus on either non system–based RA, such as prevention, audit, or training.

But what constitutes a batch? There are a number of ways to define a batch, one could argue that all calls from one switch in a day, or calls terminating within a time band, or all TAP calls in a day, or all wholesale calls in a day, and so on, represent a single batch. Or you might argue that the process is continuous, in which case there are sampling plans for that—I wanted to keep the discussion simple. In all cases, the only requirement is to make a case for a definition of a batch that you are happy to justify.

How do you take a random sample from the batch? Again there are a number of different approaches, for example, capturing every nth call that is terminated—because you have no control over the start time, duration, call type and destination and the traffic is representative of the distribution of the traffic on the switch or network, it is close enough to a random sample not to compromise the findings.

So what are the downsides?

Some organizations utilize the RA system to help with Sarbanes–Oxley compliance. I am not an expert, but my expectation would be that a process that was capable of detecting errors at the 1 in 10,000 level would probably be considered a suitable tool given the level of errors in other parts of the business.

The data is used for business intelligence reporting so needs to be 100% complete. Business decisions are not based on whether an answer is 5.01 or 4.99, or a trend in up by 1.01% or 0.99%, they are based on more significant gaps, for example, 5 or 1, or 1% not 3%, simply because the uncertainty about the external factors makes reporting to this level of detail pointless, the usual argument concerning the difference between accuracy and precision. The probability is that the sample will provide the accuracy required to make business decisions.

We want to see that all the records are being captured. RA is about balancing costs versus risks, if the increase in your operational cost exceeds the predicted value of calls missed through using sampling, then you are acting against the best interests of your business. And with the drive to lower the prices of calls, this equation moves further away from 100% sampling with time.

As yet I have not heard a convincing argument that sampling on usage is not valid, but am open to offers.

I would not elect to use sampling on nonusage data or standing data. There are few benefits to be gained primarily because the volumes involved are usually considerably less than usage data, and the rate of change is slower, so 100% reconciliation on a periodic basis works for me.

The real problem is that people are reluctant to accept sampling theory; they create individual scenarios to justify not using sampling without applying the check—how realistic is that scenario, and what would the potential cost be? It is a confidence issue, have confidence that the mathematics that has been used for many years is as valid today as it always was and be prepared to defend your position. And just to make life interesting—if you accept that sampling is the correct approach, then the argument about real-time RA disappears, which is why you are unlikely to find vendors pushing sampling.

I am not antitools; I am strongly for tools that can be easily used and are affordable to the smaller player. Using tools in a smarter manner has to be the right approach.

Final note that using test call generators is not random sampling; it is using a control sample to monitor outcomes.

A Worked Example for Using Statistics in Precision Assurance Testing

Eric Priezkalns

Most revenue assurance (RA) practitioners know little about statistics. At the same time, when asked to express an opinion on the adequacy of a size of a sample, or whether there is too much risk to rely solely on testing a sample instead of testing the whole population, most RA practitioners will give an opinion. I want to draw attention to the inconsistency this generates. Statistics is not a matter of opinion. It is a matter of fact. To know the relationship between a sample size and the confidence in the conclusion drawn from testing that sample requires the performance of calculations, not the expression of an "expert" or "experienced" opinion. It can be done by a schoolchild if armed with the right textbook and a calculator. Perhaps the saddest aspect of the deficient use of statistics in RA is that a relatively low level of statistical sophistication could be learned quite easily from standard sources, and would lead to significant improvements in the efficiency of RA practices and certainty with which findings are reached. Given the difficulty of learning about so many other aspects of RA, mastering basic statistics would be well within the capabilities of most practitioners.

To illustrate, I will present a worked example of how to apply statistics to an assurance challenge where an unusually high degree of precision is required. By so doing, I hope to show how statistical techniques can and should frame our thinking in the number of tests to perform and the conclusions that can reliably be drawn. Choosing such a demanding example should also illustrate that the more modest expectations of routine RA can easily be satisfied with much smaller samples, begging the question of how the practitioner can justify the cost of investing in systems to check an entire population in real time.

Let us suppose I wish to execute a test plan to determine how many of a certain population suffer from error, with the goal being to see if the error rate is more or less than some target ratio. There is no need to test every item in a population to reach a reliable inference about that population. It is more efficient to test a suitably selected and representative sample of the right size. This cost efficiency is most obvious when we have no idea what the result will be; if I test a thousand items without finding an error, it is difficult to justify testing a further million items "just in case." Hence sampling is particularly pertinent when first assessing the scale of problems for a population that has not been tested or measured before, and will be useful in deciding how much resources should be put into controlling and monitoring the population—whether it really does need extensive real-time monitoring across the full population, or whether an occasional sample check is sufficient.

The Challenge

For this challenge, let us suppose we have a revenue share arrangement with a content provider. Every time their content is enjoyed by one of our customers, we owe them a fee, and the size of the fee is based on the duration for which the customer enjoyed the content. The partner wants to know if they are missing revenues and can ask for an annual audit to check they are not. They are interested in errors where the duration was underrecorded. Underrecording errors mean the content provider received a smaller fee than they should. The two businesses have agreed on a contract that says if the number of errors is above a certain level, we must also pay the content provider a penalty fee. The target error ratio is 1 failure per 50,000, so we must pay the fine if the error rate is worse than this. If our own internal sample check shows that there is more than 1 failure per 50,000 then it makes sense to spend money on fixing the issues that cause the errors, to avoid paying a fine when the content provider does its audit at the year end. At less than 1 failure per 50,000 the content provider can only claim back the money on specific leakages they find during their audit, with no additional penalty fee.

A failure rate of 1 in 50,000 can be expressed as a probability of 0.00002 (which is the decimal we get when dividing 1 by 50,000). In short, we want to devise a test plan that will tell us if the probability of failure is more than 0.00002.

Confidence

Statistical confidence is as it sounds: a statement about how confident we are that a conclusion drawn is the right conclusion. When using statistics, we ask if a specific condition is met or not; the result is binary, leading to a simple yes/no answer. Hence, when drawing a conclusion, two kinds of mistakes are possible: we conclude a yes when the answer should have been no (a false positive), or we conclude a no when the answer should have been yes (a false negative). Confidence is a measure of how sure we that when we conclude "yes" the answer really is yes, or when we conclude "no" the answer really is no.

For example, suppose I want to reach a conclusion with at least 90% confidence. This means that after all the tests are performed, the chance that I reached the wrong conclusion will be less than 10%.

I can set the desired confidence level in advance, but when deciding a sample, it is not possible to know what confidence level will actually be achieved in practice. In short, whatever sample size is picked, it may be necessary to increase the sample to get the desired level of confidence, depending on what the test results are in practice. For example, suppose we sample 10 chargeable events and find an insufficient fee was passed to the content provider for all of them. If the results keep coming out the same way, we soon get a high confidence that our target error rate of 0.00002 is exceeded. On the other hand, suppose I test 50,000 events and find 1 error. That equates to exactly 0.00002 probability of failure. If the actual error rate is very close to 0.0002, I would need to do very many tests to get a high level of confidence when comparing my test results and reaching a conclusion over whether the probability of failure is more or less than 0.00002.

One final thing to note about confidence is that the confidence of not getting a false positive is different to the confidence of not getting a false negative. So when devising my test, I need to be clear about the direction of confidence I am most interested in. In the case of paying contract penalties, this will be costly and have a negative impact on the relationship with the partner, so we want to be sure that our error rate is less than 1 in 50,000. That means we are wary of a false negative—we do not want to conclude that the error rate is below the target when in fact it is above the target. We want high confidence that the rate is less than 1 in 50,000, so when the content provider does its audit, there is little chance of them being upset and demanding a contract penalty.

Had we created a different scenario, we might have wanted a high degree of confidence in the other direction; that is, we would be worried about false positives. For example, suppose a one-off investment in a new order management system becomes cost effective so long as it eliminates all errors where the current error rate is 1 order goes missing in every 1,000. In that case, we want to be very confident if we decide the error rate is more than 1 in every 1,000, because we do not want to spend money on a system that will not actually pay itself back in terms of added value. If the error rate appears to be over 1 in 1,000, but the confidence is low, it makes more sense to do more tests and see if confidence improves, rather than rushing to make an expensive one-off purchase that may not actually pay for itself in practice.

For the purposes of our challenge, let us suppose there is a very large contract penalty for missing more than 1 transaction in 50,000, so we want to be careful. We set a desired confidence level of 95%, meaning that after we have done our internal tests, there is less than a 5% chance that we decide the error rate is lower than 1 in 50,000 when it is really higher than 1 in 50,000.

A Summary of the Question and the Answer

So far, we have said we want 95% confidence that errors are less than 1 in 50,000. But we have not talked about how many tests we need to perform yet! The usual next step is to try to come up with an answer to that. That is what we need to do in practice, but bear in mind the aforementioned comments about what happens when tests results are very close to our target—we need a lot more tests to attain the desired level of confidence. We will not know if we face that problem until we start testing, so in practice we decide the sample size based on a hunch: how many errors we think we will get.

That said, suppose we want to pick the smallest possible sample size that would give 95% confidence (0.05 probability of a mistake) that errors are fewer than 1 in 50,000 (0.00002). In other words, let us suppose that we find no errors at all in our sample (because the more errors we find, the less likely we are to confidently conclude the error rate is less than 1 in 50,000). What is the sample size?

There is a theoretical minimum number of tests that must be performed to fix, at 0.05, the probability of accepting a system that has a proportion of more than 0.00002 in error. This theoretical minimum is 149,786. So if we were testing in real life, we would not design our test plans with less than this number of tests in mind. If we did plan to do fewer tests, we could never hope to conclude the error rate was less than 0.00002 with 95% confidence. Only doing the minimum number of tests leads to a very high likelihood of rejecting an acceptable system. For example, a system that in reality has an error rate of only 0.00001 would be rejected as "not good enough" 78% of the time.

This probability, the probability of rejecting an acceptable system, can be controlled. To do so requires specification of (1) the desired level of the probability of incorrectly rejecting an acceptable system, and (2) the proportion of chargeable events that fail to be processed in this acceptable system.

If the probability of rejecting an acceptable system is controlled at a reasonable level, then a sample size much larger than the 149,786 given earlier will be needed. That is, the 149,786 should be taken very much as a minimal value, and in practice one would expect to use larger values. Only if we expected the test results to be near perfect; that is, we never find an error in practice, would we execute a test plan based on the minimum number of tests.

The calculations on which the aforementioned figures are based assume that the chargeable events chosen for testing are selected at random from the population of events generated over a given period of time.

Calculations

Let E_m be the measured duration of the event, and E_t the true duration. There is a deemed to be an error in the fees if

$$P(E_m < E_t - \delta) < 0.00002 \tag{10.1}$$

where

- P(A) means the proportion of chargeable events in the entire population that satisfy condition A.
- δ is a tolerance relating to the accuracy of measurement; that is, if a tolerance of 2 seconds is set, the recorded duration is not considered in error if it is 1 second less than the actual duration.

Here, the term "population" refers to the population of events over a given period of time. The population may be homogeneous or heterogeneous, but the calculations refer to the entire population. Throughout, we assume that the sample of events chosen for testing are selected at random from the entire population of events. This example is solely concerned with the event's duration, though the same statistical principles and arguments apply to any property of events where the aim is to determine whether a certain proportion of events lie in a given interval.

Let us say that $P(E_m < E_t - \delta) = \theta$. Then 10.1 requires $\theta < 0.00002$. To test this, we will take a sample of n events, and estimate θ from θ', the proportion of the n sampled events that have $E_m < E_t - \delta$.

Since the aim of this testing is to protect the relationship with a contractual partner, we need to be confident that we are right if and when we assert that $\theta < 0.00002$.

When determining a confidence level it seems clear that when, in fact, $\theta \geq 0.00002$ we wish to have a low probability of wrongly concluding that $\theta < 0.00002$. We will aim to restrict the probability to 0.05 or less.

Unfortunately, these conditions alone are not sufficient to determine an appropriate minimum sample size. This is easily seen by considering a scenario in which the true θ is infinitesimally larger than 0.00002. Now, no matter how large a sample of calls is taken, almost 50% of such samples will have the proportion θ' less than 0.00002. That is, almost half the time when we observe a proportion less than 0.00002 and conclude that the system is acceptable, it will not be. No matter how large a sample size we take, we can never reduce to less than 5% the proportion of such samples that have θ' less than 0.00002. To define the problem sufficiently precisely that a minimum necessary sample size can be determined, some additional constraint must be applied.

To overcome the difficulty illustrated in the preceding paragraph, we need to fix a threshold, less than 0.00002. Call this threshold f. Now, for a given sample size, an elementary calculation can be made to determine the probability of observing $\theta' \leq f$ when $\theta = 0.00002$. (Such calculations are illustrated later.) Call this probability β. Then, if we observe a value of $\theta' \leq f$, either $\theta \geq 0.00002$ and an event of probability less than or equal to β has occurred or $\theta < 0.00002$. In particular if $\beta = 0.05$ then we will only conclude that $\theta < 0.00002$ when in fact $\theta \geq 0.00002$ less than 5% of the time. This is what is required.

For fixed threshold f, the sample size n and the value of β are inversely related, so that we can choose n to make $\beta = 0.05$.

The distribution of θ' is *binomial* (n,θ) and a standard approach to the calculations in such problems is to use a normal approximation. This yields the probability of observing $\theta' \leq f$ when $\theta = 0.00002$ with a sample size of n to be

$$\Phi\left(\frac{f - 0.00002}{\sqrt{0.00002 \times 0.99998 / n}}\right).$$

We will set this to 0.05, so any pair of values (n,f) that satisfies

$$\Phi\left(\frac{f - 0.00002\sqrt{n}}{0.004472}\right) = 0.05$$

will do. The 0.05 point of a standard normal distribution is -1.6449, so we require the (n,f) pair to satisfy

$$n = \frac{0.00005411}{(f - 0.00002)^2} \tag{10.2}$$

Table 10.1 gives (*n*,*f*) pairs satisfying the 95% confidence level requirements, assuming normal approximations.

However, with such small values for θ we need to be wary of the adequacy of the normal approximation. This is partly because the values of θ′ are bounded below by 0. With modern computers, such approximations are unnecessary, since exact binomial calculations can be performed. The calculations that follow use the Splus *binom* function.

The minimum sample size will be attained by adopting *f* = 0. This can be seen from the normal approximation in 10.2 since *f* ∈ [0, 0.00002] but also follows from the fact that the *f* corresponding to the 5% quantile of the binomial distribution will increase as *n* increases. If *f* is set to 0, then using the exact binomial calculations, a sample size of 149,786 is required to give β = 0.05. In other words, with a sample of 149,786 events, if we observe no errors in the sample then the probability of concluding θ < 0.00002 when in fact θ ≥ 0.00002 is only 0.05.

Although the minimum sample size is given by adopting *f* = 0 this is not a good idea because it leads to a very low power; there is a high probability of concluding the error rate exceeds the required 0.00002 when in fact it is lower. Suppose that θ = 0.00001, which is half the error rate needed to trigger a contractual penalty. If tested using a minimum sample size, then the probability of wrongly concluding θ ≥ 0.00002 is given by the probability of observing one or more errors in the sample, with a binomial distribution of 149.786 and parameter θ = 0.00001. This probability is 0.7764. That means there is a greater than 77% chance of wrongly concluding that the required error rate is exceeded if using the "minimum" sample size when the error rate is in fact half of the stipulated limit.

For another example using the minimum sample size, the probability of incorrectly concluding that θ ≥ 0.00002 (=2 × 10^{-5}) when in fact θ = 4.6 × 10^{-6} (a much lower error rate) is still 0.5.

Although *f* = 0 gives the minimum sample size required to control the probability of incorrectly deciding that the system is acceptable when it is not, this still leaves a high risk of rejecting a system that is well within the required performance limits. To decrease this risk, we need to increase the sample size and increase *f*. For any given *f* ∈ [0, 0.00002], binomial calculations like those above can be made to determine the sample size that will control the probability of accepting an unacceptable system at the required 5%. For example, if *f* = 0.00001, a sample size of 387,670 is required to give β = 0.05. This means we will conclude that the system is within the required accuracy bound only if we observe less than 0.00001 × 387670 ≈ 4 errors

Table 10.1   Example Sample Sizes

f	*n*
0.000005	240,489
0.000010	541,100
0.000015	2,164,400

in the sample. With this sample size, the probability of rejecting a system where $\theta = 0.00001$ is 0.5, lower than the 0.78 given earlier.

Choosing the Decision Threshold

We need some rationale for our choice of f. The probability of incorrectly rejecting an acceptable system α should be part of this. For example, we might want to set a limit of $\alpha = 0.2$. Specifying the probability of incorrectly rejecting a perfectly acceptable system is not sufficient to determine f and n. We also need to specify the performance of the "acceptable" system for which the rejection probability is being controlled at 0.2.

If we believe our system has an error rate of p_1 (less than 0.00002), then we can choose a unique value of f such that the following conditions are all satisfied

(a) $p_1 < f < 0.00002$
(b) $P(\theta' < f \mid \theta = 0.00002) \leq \beta$ (set to 0.05 above)
(c) $P(\theta' > f \mid \theta = p_1) \leq \alpha$

This requires specification of α as well as of p_1. Denoting the cumulative binomial distribution with sample size n and parameter p by $pbinom(f; n,p)$, the relationships

$$pbinom(f; n, p_1) = 1 - \alpha \qquad (10.3)$$

and

$$pbinom(f; n, 0.00002) = \beta \qquad (10.4)$$

each provides equations that relate f and n. Solving these simultaneous equations gives values for f and n.

Is Revenue Assurance CAT or CAVT?

Eric Priezkalns

It is nothing to be proud about, but I once had a flaming row with a revenue assurance (RA) expert who works in the Far East about whether the objective of RA is completeness, accuracy, and timeliness (CAT) or completeness, accuracy, validity, and timeliness (CAVT). I was giving him a lift in my car at the time, and the row got so bad that I had to pull over and let him out. Now maybe the ordinary RA practitioner may not get quite so excited about this debate, but we both got very worked up. In short, I thought, and still think, RA is CAVT and he argued it is just CAT. In fact, he went further and said validity was implied by accuracy. That really wound me up, as I dedicated three years of very hard and boring work to accountancy studies which insisted that accuracy and validity are two completely different concepts. So at the conference I chaired last week, I was forced to smile through gritted teeth when one speaker gave his explanation of RA, mentioning CAT but failing to talk about validity.

Where am I going with this? Well, I think the CAT definition of RA is ambiguous, open to misunderstanding, and either logically sloppy or ethically unsound. To explain why, we should begin by defining the C, A, V, and T in CAVT.

- Completeness: There is a data record for every event or object in the real world.
- Accuracy: Data records accurately describe the event or object in the real world that they correspond to.
- Validity: Every data record corresponds to an event or object in the real world.
- Timeliness: All data records are captured and processed on a timely basis.

Using these definitions, the difference between CAT and CAVT is very straightforward. Data records that are CAT need not be valid. For every event or object there is a record, and that record is accurate and is processed in a timely fashion, but there may also be other records that do not correspond to any real event or object. Rather obviously, this is just another way of saying you may have some invalid records in your data. For RA practitioners who have had accountancy training, the distinction will seem natural. So arguing for a CAT definition of RA would mean arguing that it is not the responsibility of RA to find invalid records. That would mean RA departments would not try to identify duplicate records, or charges for calls that never took place, or bills for equipment that were never provided to the customer, or inventories that say network has been provisioned when it has not. I have been in the room when RA practitioners have argued for this strict CAT view of RA. Everybody is entitled to their opinion, but I think this view of RA is plain wrong. It is wrong on three counts: it is inefficient for the business, weak for cost management, and inadequate to support good corporate ethics.

Testing for validity is the inverse to testing for completeness. If the RA department is going to test for completeness, it is more efficient for the business that the department tests for validity at the same time rather than relying on another department to do it. Invalid records are often a sign of poor cost management. While invalid records may lead to higher revenues in the short term, if customers find out they are being overbilled then they are likely to lead to lower real revenues in the longer term, as a result of credits and churn. Invalid records also lead to increased costs when you consider the staff cost involved in handling and resolving customer complaints. Finally, ignoring evidence of invalid records, if this resulted in overbilling, would be a serious ethical breach. I hence cannot agree that RA should only assure the CAT of revenue processing, but ignore validity.

To be fair to the former interlocutor who was ejected from my car, he was not arguing that RA did not need to address validity. His argument was semantic. He thought that there was no need to use the word "validity" because validity is already implied by accuracy. In his view, a record could not be accurate if it was not also valid. This is the same as saying his CAT is identical with my CAVT. I dislike this argument for a number of reasons.

First, I do not see what is gained by taking two useful and distinct definitions for the words "validity" and "accuracy" and insisting that accuracy covers both. It is perfectly possible to construct different tests for validity and accuracy. It may not be possible to construct a test that addresses both at the same time. So, as an auditing principle, it is helpful to keep them distinct to ensure you really did check them both. For example, if I can reconcile that there is a CDR for every call made, then all the CDRs are valid, but this does not prove that the details in the fields of the CDR are all correct. Because testing all the details of all the CDRs would be a prohibitive workload, it would make more sense that I check a sample of CDRs to assess whether fields are accurately populated. Similarly, if I found there were some invalid CDRs in my check for validity, I would not waste time including them in my sample test of accuracy. It is also possible to make mistakes if people are not clear on their purpose. For example, in one tax dispute I saw an RA practitioner check if every billed account in the billing system resulted in an update to the ledgers in the accounting system. He then compared the details of the records to ensure they were consistent. He found lots of examples where a bill had been issued, but the ledgers had not updated, and also of cases where the ledgers had been updated incorrectly. The results were explained to the taxman. The taxman was not impressed. The practitioner had traced forward from billing to ledger, and checked accuracy whenever there was a match. It had never occurred to the RA practitioner to also trace backward from ledger to billing system. In this case, there were also ledger entries not supported by the billing system, that is, invalid ledger entries. The practitioner had got so excited by the detail of checking accuracy, as he understood it, that he never thought to do a proper check for validity.

The second reason I dislike saying accuracy includes validity is that some people honestly believe RA does not include validity as an objective. That does not mean they do not believe it necessary to check for accuracy. While I disagree with their point of view, I have to permit them some way of expressing their point of view. With the CAVT definitions used earlier, it is possible to describe their point of view, because they simply do not agree in including validity as an objective of RA. However, if accuracy somehow implies validity, then it becomes very difficult to describe their point of view. You end up needing to say they believe the objectives are CAT but excluding validity, which just takes us round in a circle given that the whole point of dropping the "V" was because of an insistence that validity be considered an integral part of accuracy.

My interlocutor was very insistent that he had been trained as a computer scientist and that in computer science validity is an element of accuracy. From his perspective, a record is not accurate if it does not describe a real thing. He argued that the accountants had overcomplicated a situation that the computer scientists were keeping simple. As somebody who spent a semester teaching mathematical logic to computer scientists, I did not agree with his opinions. If logic is the basis of computer science, then it is possible to describe, in logic, both my preferred definitions for validity and accuracy, and his preferred definition of a kind of accuracy that also includes validity. In logic, no definition is superior to any other. In logic, there is no conflict; it is only when we describe the logical rules in English that a clash may occur. The usefulness of a rule is determined in practice, not through logic. But there is no advantage to the computer scientist in trying to curtail the logical basis for the science, any more than it would be an advantage to build a computer that could turn bits from 1 to 0 but could not turn them from 0 to 1. For those of you interested in the logical foundations, I have constructed some definitions of the CAVT objectives, and of my interlocutor's alternate definition for accuracy, including validity, using the notation of predicate calculus. I turn to predicate calculus as a formal logical system well suited to the task; for those wanting an introduction to predicate calculus, I recommend the *forall x* by P.D. Magnus (Magnus 2005), which is available for free under a creative commons license from his Web site, fecundity.com.

Preliminary Definitions

Define a function "Data-record": Data-record (A) is true if and only if A is a data record.

Define a function "Real-world": Real-world (A) is true if and only if A is an event or object in the real world.

Define a function "Record-of": Record-of (A, B) is true if and only if Data-record (A) is true, Real-world (B) is true, and A uniquely corresponds to B.

Define a function "Consistent-with": Consistent-with (A, B) is true if and only if Record-of (A, B) is true, and all the details of Data-record (A) are consistent with the facts about event or object B.

Define a function "Timely": Timely (A, P) is true if and only if Data-record (A) is true, and A is created and processed before the deadline or deadlines stated in policy P.

To aid readability, variable x will always represent data records, and variable y will always represent events or objects in the real world. In other words, if we defined functions to clarify whether something is a data record or it is something in the real world, we would say Data-record (x) is true and Real-world (y) is true.

Definitions of Revenue Assurance Objectives
Using Predicate Calculus

Completeness: $\forall y\ \exists x$ (Record-of (x,y))
> In English, this would be described as "for all y, there is an x, such that x is a record of y."

Validity: $\forall x\ \exists y$ (Record-of (x,y))
> In English, this would be described as "for all x, there is a y, such that x is a record of y."

Accuracy: $\forall x\ \forall y$ (Record-of $(x,y) \rightarrow$ Consistent-with (x,y))
> In English, this would be described as "for all x, and for all y, if x is a record of y then x is consistent with y."

Timeliness: $\forall x$ (Timely (x,P))
> In English, this would be described as "for all x, data record x is created and processed before the deadline stated in policy P."

Alternative form of "accuracy" rule which also delivers validity: $\forall x\ \exists y$ (Consistent-with (x,y))
> In English, this would be described as "for all x, there is a y, where x is consistent with y."

One thing is for certain. Whenever you go to a bigger RA conference, you will probably find at least one speaker who describes the objectives of RA using CAVT, and another who describes it using CAT. I do not think I am biased when I say that CAVT probably gets mentioned slightly more often than CAT, but there is not much in it. Certainly I am not alone in preferring CAVT. For example, a CAVT definition of RA is given by the vendor Cartesian in the common terminology for RA they made available for free download from their web site (Cartesian 2010). However, I do hear plenty of people who describe RA in terms of CAT with no mention of validity. Whoever is right, the reason I raise the question is more than just because I think the argument is interesting or important. Some people have started reaching the conclusion that the practice of RA is mature in an increasing number of communication providers and is ready to evolve to take on new challenges. I question that. For RA to be mature, its foundations must be secure. Understanding whether the objectives of RA are CAVT or CAT is fundamental to determining the scope and purpose of RA. It is a principle that needs to be established and consistently communicated to everybody working in that team before rushing onward to expand the horizons. I have borrowed from logic because the debate and conclusions need to be scientific, not just semantic. If RA really is going to mature as a discipline, it will have to pass tests like whether the practitioners understand the difference between CAVT and CAT, and have come to a consensus on which is the correct expression of objectives. For those of you reading this, I recommend you reach your own conclusions on CAVT versus CAT, before moving on too quickly. At the very least, be clear about your conclusions before you ask for a lift in my car ...

North American Revenue Assurance: Got Your Number?

David Leshem

At some point of time I had noticed that I do not find too many contributors to *talkRA* from North America. Even though the readership is there, we lack contributions. American society has contributed greatly to the world economies and culture, with myriad inventions addressing every aspect of our lives and culture, not forgetting the infamous French fries or spaghetti bolognese (any attempt to mention this latter dish at a common Bologna restaurant would risk the guest being thrown out by the chef).

So, back to North America and RA.

The North American numbering plan, where all phone numbers, both mobile and fixed, use the same numbering system imposes a unique definition on how business is conducted.

The fact that CSPs in the rest of the world (another fine American-invented phrase) use unique nongeo prefixes for mobile phones makes life rather different when one tries to compare the industry challenges with the North American numbering plan and the implications for RA.

In North America, the called party (MPP) has to pay for the call. The fact that in the United States and Canada it is impossible to tell whether the called party is a landline or mobile imposes an MPP regime, the opposite to the CPP convention where the caller pays. In some cases the providers do offer a special ring tone when calling mobiles. In my eyes, this is a limited remedy.

As a consequence of MPP, the chargeable rate is determined by a myriad of tables and call jurisdictions. These charges are based on NPA-NXX tables and zones to determine whether the call has attributes (x,y) where x = {inter-LATA, intra-LATA}, y = {inter-State, intra-State}. On top there are "corridors" that override the above jurisdiction and also metro areas and more. For simplicity, I will not go into "1-800," and related numbers.

Needless to say, this complexity also affects interconnect and other settlements between carriers that also must follow this uniquely North American approach. In my view, if the RA sector outside of North America focuses on whether the OSS systems are correctly performing in view of recording, conveyance, charging, and provisioning, the North Americans could instead focus greatly on "guiding" rules that are complex and where one wrong or out-of-date attribute in a table could cause a lot of harm. Let me illustrate how complex the situation can be. An average reference table system for a US telco would be comprised of say tens of tables: NPA-NXX and small NPA-NXX, CLLI codes tables, and more. Vendors like CCMI and Valuecom make a living from providing such tables on a monthly subscription basis to various TEM vendors and telecom consultants.

The rates might look simple; however, the guiding logic, which involves guiding to a customer and guiding to a service, is rather complex. This is the driver of the business models of vendors like TEOCO and ATS, amongst others. As an anecdote of how things can be interesting, some rates involve distance calculation. There are at least two formulas for distance calculation. The most accurate one is based on NPA-NXX V&H coordinates and calculates using the square root. The alternative formula by AT&T is based upon AT&T V&H figures and an iterative approximation. This alternative exists because, back in 1948, computers could not calculate square root along with correction logic.

The North American market is large enough to keep an admirable vendor like TEOCO doing well. After all, for a midsize company, there is so much business in the United States that sometimes it is not worth showing your cards just to play outside the United States.

Since the domains of rating and billing verification, for retail, wholesale, and interconnect, fall within the pillars of the RA practice, it is little wonder that a generic solution would fit both the North American and rest of the world markets. If one would say that in the United States you can get "all you can eat" price plans for a fixed amount so why worry about the issues above, my reply would be (1) read the fine print and ask what does the "all" stand for, and (2) interconnect is the major revenue stream and rating is still required anyhow.

To summarize my points, RA in North America has rather different aspects and challenges to RA in the rest of the world. The *TMF RA Guidebook* is generic enough not to be concerned with such differences. As for me, I am curious. On the other hand, quite a few American vendors view CSPs in the RoW with a perspective more suitable to the North American market. I recall an immortal comment by the American CEO of a large solution vendor. He said to me once at some telco conference: " … these Europeans are not all the same … " I replied " … well the UK is not even part of the continent anyhow."

Theory of Symmetrical Error

Eric Priezkalns

Apologies, faithful readers. This is one of my most common rants being recast into yet another guise. However, I felt the need to do it because I think the idea is sound, but I have no convenient label for it. That means I have to explain the idea every time I want to discuss it. Forgive me, but if I write the idea down, and give it a name, at least I can use the name in future and point to this article as an explanation of what the idea is. So what is the idea? Well, for want of a better name, let us call it *the theory of symmetrical error*.

My experience suggests that the likelihood of errors in complicated systems, be it human error, system error, errors in data, or however you want to categorize errors, is not well correlated to the consequences of those errors. Errors that are easy to commit may have significant consequences. Errors that are highly unorthodox and unlikely to be repeated may have negligible consequences, though they are errors all the same. If leakage is the product of a mistake, and not caused deliberately—so I will exclude the impact of fraudulent activity from this analysis—then I believe those mistakes are no more likely to reduce profits than they are to increase them. This symmetry of consequences, and the equal likelihood of consequences on either side, is what I mean by symmetrical error.

Of course, an external party may be vigilantly looking for errors that negatively impact them, to protect their own interests. However, the likelihood of discovering the error is not related to the likelihood of committing the error, so this is irrelevant to the point. All the theory concerns is the likelihood of an error being committed, not the likelihood of it being found or how hard people will look for it. The probability of committing an error in one direction, and the distribution of the magnitudes of the errors in that direction, is approximately equal to the probability and distribution of magnitudes for errors in the opposite direction. It is approximate because I concede there must be some limits to the symmetry. For example, if a bill should really be stated to be for x, it is easier to imagine an error where it is instead stated as $2x$, $3x$, $4x$, and so forth, and hence x, $2x$, $3x$ over the correct amount, than it is to imagine it is stated as 0, $-x$, and $-2x$ and hence x, $2x$, and $3x$ under the correct amount. Similarly, data may be "lost" at many stages, but may be wrongly created or added to (duplication aside) less often, so there may be some kind of one-sided errors that skew the overall results.

If the theory of symmetrical error is approximately correct, then it has some important conclusions for revenue assurance (RA). First and foremost, if error is symmetrical then RA, if conducted in an unbiased way, is not an activity that can promise to regularly generate a positive return for the business based solely on the value of the errors it finds and corrects. Measured on this simple basis, if error is symmetrical, then RA is as likely to reduce business profits as to increase them. The only way to consistently increase profits would be to selectively detect, investigate, or resolve errors in one direction, and ignore errors in the other direction. One could assume that external parties will protect their own interests, but in this era where corporate governance is on the lips of so many executives, it might be considered foolish to protect your own interests and play fast and loose with everyone else's. Second, promising skewed results leads to skewed expectations, skewed collection of data, skewed interpretation of that data, and skewed remedial action. In other words, the prophesy of generating financial benefits from RA is self-fulfilling. If you want to generate positive results, only acknowledge errors that will generate a positive return when resolved; ignore all other errors. This is unhealthy if the interests of customers and partners deserve to be given more than lip service. Third, protecting the interests of the company and protecting the interests of external parties is not fundamentally different. They are both addressed by the same kinds of controls, monitoring and preventative activities, because the errors themselves are alike more often than not. The only reason for separating responsibility for executing controls to protect the business from controls to protect external parties is the difficulty, or undesirability, of motivating staff to do both at the same time. Finally, estimates of error fall between two extremes postulated by RA practitioners. At one extreme, when estimating errors that reduce profits, the average level of error is grossly exaggerated. At the other extreme, when estimating errors that increase profits, because they wrongly overcharge customers, the average level of error is understated by the numbers in the public domain.

I will make two observations to support the theory of symmetrical error:

1. Estimates of errors for retail transactions show the greatest divergence depending on the direction. Estimates of revenue loss due to failure to charge customers for transactions, or undercharging of transactions, are high, while estimates of the overcharging of customers are low. This result suggests that errors are highly asymmetrical. However, examination of the specific basis used to calculate errors in each direction highlights severe inconsistencies in approach, each selective of what data is used and how it is interpreted. Skewed mechanisms to calculate error suggests underlying error rates are more similar than the headline error rates in each direction.

2. When settling charges between communication providers, whether interconnect, roaming, or other wholesale intercarrier charges, any errors in the final settlement must be symmetrical. Whoever "loses," the other party "wins" by the same amount. It is worth noting that many commentaries on RA discuss the risk of leakage from intercarrier transactions and treat it as comparable to other kinds of risk of revenue leakage, but fail to highlight the unequivocal zero-sum nature of any errors in these transactions. At the same time, typical discrepancies found when reconciling interconnect or other intercarrier charges are significantly less than the reported industry survey estimates of leakage, while significantly higher than self-reports and regulatory compliance reports of overcharging of retail customers.

The symmetry of error is not a trivial problem for RA. If the conclusions are correct, then the RA profession needs to put its house into order, and admit that it has an equal job to do in both protecting the business and its customers. If the business is at risk, then customers are at risk. If the customers are not at risk, then the business is not at risk. If the conclusions are incorrect, then there is still a job of work to do, in explaining why error is so asymmetrical. Asymmetry should not be assumed; it needs to be shown using consistent data and consistent interpretations and calculations based upon that data. And if asymmetry can be shown, there is still a responsibility to explain why risk is so significantly skewed in one direction. The need for an explanation is made all the more pertinent by the fact that asymmetric error rates would be remarkably convenient for practitioners of RA. Asymmetric errors reduce the burden for protecting customers and guarantee overall financial returns for the business. Any good scientist should be wary of theories that give results that also coincide perfectly with their own interests. Because asymmetry would be so very convenient, RA needs to apply a higher standard if it is to conclude that errors tend to be asymmetric.

This is the theory of symmetrical error, if you will. I know it is not a very scientific exposition, but then RA is not a very scientific discipline, as performed at present. Sciences need to start somewhere. The theory of symmetrical error would be a good place to start if RA is going to move on from making promises to explaining how they are kept.

Research Found Revenue Assurance to Be Responsible for Insanity

Güera Romo

It is with utter frustration that I write this. I feel like saddling up my transformation horse and changing the archaic processes of academic institutions. It is perhaps not their fault.

For my university research, I am reading the same revenue assurance (RA) material over and over in different forms. Some industry magazines have articles that portray a reality closer to our experience within RA. However, very few make reference to where they got such insight. Is this the author's experience, his opinion or somebody else's opinion he happens to capture aptly in an article? It is rather frustrating when you read a sentence and realize you have read the same exact sentence before.

Now consider this for context first: I am using over 40 sources for one chapter of literature review. For the essence of RA, I am working through more than 40 sources, which range from a 2-page industry magazine article to a 705-page book on revenue management. Trying to remember the context of each of these sources so that it can be consulted again in subsequent sections is daunting on its own. Trying to figure out which source is more authentic than the next is quite difficult.

What is driving me to consume large quantities of chocolate at midnight? The fact that the real goodies do not appear in literature suitable for academic research. The gap between the reality on the floor and half the "how to"s does not allow me to portray a vision of RA for the future. In a sense it is almost a case of "whoopee, somebody integrated 12 years of literature to serve as *RA Over the Last Decade for Dummies*" because it certainly does not provide much of a basis to build on other than confirming that we RA practitioners are all over the damn show. But then again, those of us interested enough in the discipline know that.

The upside is that there are more than 40 sources to use, and even more still if I really wanted to incorporate them all but that would not add any value. It is just sad that good material is available in blogs and speaking to people too busy in the trench to bother with approaching an academic journal to publish such knowledge. That is, tacit knowledge that you can never give justice to in a 2-page opinion-flavored article.

Perhaps we should hunt down the review committees of academic journals in the ITC or telecom fields and offer them some of my chocolate in exchange for making the process easier to publish real stuff.

An Entity-Relationship Model for Revenue Assurance Controls

Eric Priezkalns

This article introduces a novel idea in revenue assurance (RA): the use of an entity-relationship model to describe what RA is and what it does. The reasons for using such a model are threefold:

1. An entity-relationship model is precise, leaving no room for ambiguity.
2. The entity-relationship model can be interpreted as a description of what information should be documented as part of the RA process. This documentation can be considered a deliverable of any process and hence a way to describe a process in terms of tangible and verifiable deliverables.
3. Entity-relationship models are typically described to model databases. A generic model for RA controls could literally be instantiated as a database and this database could be used to represent and document the practical work of a real RA function.

Before we go further, let me briefly explain what an entity-relationship model is, to those unfamiliar with the concept. An entity-relationship model is an abstract and conceptual representation of data. Entity-relationship modeling is typically used to produce a schema or semantic data model of a system, often a relational database, and its requirements, in a top-down fashion. Models are often presented diagrammatically. Entities can be thought of as "things" or objects that exist in the real world. Two entities may have a relationship to each other. "Oslo is the capital of Norway" hence can be thought of as a statement that there is a "capital" relationship between the entity that is the city of Oslo and the entity that is the country Norway.

Relationships have cardinality, which is another way of saying that there are rules about the number of things on each side of a certain kind of relationship. Each country has one and only one capital, so the cardinality of that kind of relationship is a strict one-to-one. A country may have zero, one, or more borders with other countries, so the cardinality of that relation is one to zero, one, or more.

The relationship between a country and a continent is one to one or more, because every country is part of at least one continent but some, like Turkey, straddle more than one. In some countries, the head of state is a president, but not all; the cardinality of the relationship between countries and presidents is hence one to zero or one. The inhabitants of a country will have an address; several people may share the same address and some people may have more than one address, giving rise to a many-to-many relationship between citizens and addresses.

Now let us take a look at Figure 10.1 for the entity-relationship model for RA controls, and walk through what it means.

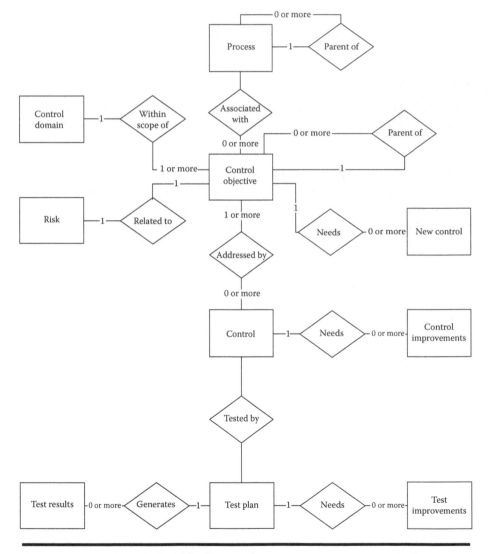

Figure 10.1 Entity-relationship diagram for revenue assurance controls.

In the figure, rectangles are entities and diamonds are the relationships between them. The cardinality of relationships is shown by the numbers and words on the lines between entities and relationships.

Let us begin with the top of the diagram, and the process entity. Of course, communication providers have many processes. Processes may also be described at a greater or lesser level of detail, so the dunning process would be considered a subset of the billing and collections process, which is in turn a subset of the process of managing the relationship with the customer. This sense of a process hierarchy will be familiar to anyone who has used the TM Forum's enhanced telecom operations map (eTOM). The sense that one process may be a subset of another process is captured in the diagram by the relationship "parent of" that creates a loop out of and back into the entity "process." The cardinality of this relationship is that a process can only have one parent process but a process can have zero or more children. This simple construct is included in the entity-relationship model to enable the RA practitioner to zoom in and out of process detail, as appropriate for the specificity of control objectives and controls they are interested in. To follow our previous example, the RA practitioner might be taking an interest in a control objective that relates solely to dunning, or to the whole of billing and collections, or to the entirety of managing the relationship to customers.

Going down from the process entity, each process may have zero or more associated control objectives. Each control objective relates to one and only one process. A control objective is an RA objective that should ideally be (but may not be) fulfilled in relation to a process.

In the same way that processes may be the parents or children of other processes, so control objectives may be the parents or children of other control objectives. Hence, if accurate rating is a control objective for the rating process, then accurate input of rating reference data may be a control objective for the subprocess of maintaining rating reference data.

From a practical perspective we need to group the work of RA into manageable "chunks" or domains. Unless we have a small and simple communication provider, the work of RA needs to be split between several people, and in many cases we will also want to divide the work done by staff to separately manage and report on progress for natural subgroups of their work. We could do this by using the capacity to have parent processes, and hence break down all RA according to process, but this may sometimes be an inconvenient way to organize work. A control domain most naturally relates to what some practitioners call "streams"; this may be a division between prepaid and postpaid, between wholesale and retail, between the product lines of a multiplay provider, and may indeed be combinations of these. By keeping control domains separate to our description of processes, it makes it easier to recognize that there are practical (and somewhat arbitrary) decisions on how to handle processes that could be said to straddle "streams" (in the way that roaming, voice, and VAS services may be covered separately by the RA team, but they all come together with common processes when presented on the customer's bill).

For precision, this model enforces a one-to-one relationship between risks and control objectives. The purpose in doing this is to also describe risks at the same level and with the same specificity as control objectives. Risks are perceived to be the mirror of control objectives. If a control objective is a good thing we want to make sure happens, then the risk is the possibility that a bad thing happens instead.

The aim of RA should be that controls are implemented to deliver the control objectives. A control objective may be addressed by zero, one, or more controls in actual effect. Any given control may satisfy one or more control objectives. As the RA practitioner does his or her work, gaps may be identified where there are control objectives not addressed by controls. Where no control exists, but the RA practitioner designs a control that could be implemented to fill a gap, this can be captured with the "new control" entity. This entity refers to designs for controls that have not been put into effect (controls that do exist are covered by the "control" entity). A control objective may need zero, one, or more new controls. Conversely, a new control is prompted to address a single, specific control objective. Though controls can address more than one control objective when implemented, the enforcement of a single control objective per new control is chosen to ensure that the design is assessed and measured relating to the efficacy for one specific objective.

Where a control exists to address a control objective, but the RA practitioner has identified ways it can be improved (to be more effective or more efficient) this is captured using the "control improvements" entity. Any control may need zero, one, or more control improvements. A control improvement relates to a single, specific control. Like the "new control" entity, the "control improvement" entity is a "to do" list of work that the RA practitioner would like to see performed to reduce risks and leakages.

To identify if a control is working effectively and efficiently, the RA practitioner will need to test the control. This begins with a test plan, and there is a one-to-one relationship between a control and its test plan. The test plan generates zero or more test results (zero if the tests are never actually performed, more if the test plan has been executed repeatedly). The "test results" entity is effectively an archive for all results from each time the test plan has been executed. Executing a test plan more than once is relevant if there are questions about how changing circumstances might make a control less effective or relevant over time, or just to show continuity of control effectiveness, as might be pertinent to demonstrating control adequacy to external parties like auditors. Finally, the RA practitioner's review of the test plans or of test results might highlight deficiencies with the test plan as well as deficiencies with the control being tested. Where a test plan needs to be rewritten, the "test improvements" entity captures the zero or more improvements needed to the test plan for when it is next executed.

Summary

This entity-relationship model can be implemented as a database. If instantiated as a database, it structures and documents the nature of the work done by the RA department. It does not organize the raw data and results from checks of OSS/BSS systems, but it does explain what checks there are and how they relate to the processes of the communication provider, the other checks in place, the objectives of RA and the risks it encounters, and the way the RA practitioner uses reviews and tests of controls to drive improvement. Even without the database, understanding the conceptual relations provides a precise way for the RA practitioner to structure their work and methodically formulate what progress they have made and what they still need to do.

Transdisciplinary Approach to Problem Solving

Güera Romo

Michael Crow, President of Arizona State University is striving to create the campus of the future, according to Newsweek Magazine (Theil 2008). He is approaching the running of the school like a CEO would run a business. What is striking from this article is the abolishment of traditional departments and combining these functions into what are called "transdisciplinary" institutes. His objective is to get experts in different fields to work together to solve problems by thinking outside their disciplines. For example, the new School of Sustainability features professors from 35 disciplines! Can revenue assurance (RA) learn from what Michael Crow is doing?

I recently had a discussion with a local university and a similar concept was discussed. While some universities still have very strict enrolment rules per department or course, these are normally more related to preventing an administration nightmare, than truly ensuring that the candidate has adequate foundation knowledge to progress in his/her new field of study. There are ways to navigate through these lines of study that can take a student from one field of graduate study to another field for postgraduate study. Is this regarded then as interdisciplinary, multidisciplinary, cross-disciplinary, or transdisciplinary?

- "Interdisciplinary" refers to professions where the traditional academic boundaries are crossed and its goal is to connect or integrate the knowledge from the different disciplines to establish knowledge in a new field of study such as in the case of geobiology.
- "Multidisciplinary" refers to joining the knowledge from different disciplines without integration. The aim here is not to establish something new but to look at the same reality from different perspectives to explain the phenomena.
- "Cross-disciplinary" discusses a subject in terms of another. It looks for metaphors or likenesses to explain a concept or object in another language if you will, one that would be understood by someone trained in another field of study or discipline.
- "Transdisciplinary" uses the multidisciplinary approach with the difference of engaging these disciplines as stakeholders in solving a problem. It is not only

viewing the reality from different perspectives to explain why the reality is such but is actually working together to overcome a problem without integrating the knowledge in each of the individual disciplines.

Given the aforementioned descriptions, I would say that "transdisciplinary" best describes the ideal approach to RA.

Minimize the Risks with Voice over WiMAX

Eric Priezkalns

WiMAX is both an exciting way for new competitors to disrupt the market for Internet access, and an alternative delivery mechanism for established players. With worldwide interoperability for microwave access (WiMAX) comes voice over WiMAX (VoWiMAX), another disruptive force that further threatens to hasten the decline of traditional circuit-switched voice products. As you might expect, supplying WiMAX demands a revision of the communication provider's methods for managing and assuring revenues. Communication providers may lack sufficient data to perform revenue assurance (RA) checks in the traditional sense. Meanwhile, a competitor or fraudster may exploit the WiMAX network to provide their own voice services, making use of the communication provider's (CP's) own network to undermine their revenues.

WiMAX providers should specifically extend their RA processes. To start with, communication providers need to treat softswitches like any other switch, by performing independent assurance checks on the completeness, accuracy, and validity of the records the softswitches produce. At the same time, they need to monitor third-party VoIP traffic being carried by the WiMAX network. Fraud protocols need to be upgraded for the new specific risks introduced. The WiMAX offering needs to be integrated into the IP-ready monitoring capability. Voice over WiMAX needs to be specifically targeted within the context of an overall strategy for assurance of IP-based services. CPs need to separately measure the relevant traffic, even if this needs to be calculated by extrapolating from a stochastic sampling process based on IP probes.

There will be some cannibalization of voice revenues, so the trick is to both forecast how much cannibalization will take place, and then measure the actual degree to which it takes place. Tackle the issue not by pretending it will go away, but by promoting packages of services to tackle it head-on. For example, if cannibalization is high, one option is to offer "eat-all-you-want" services for the WiMAX network that emphasize the benefits of using it for voice traffic, but focuses on ensuring those services are provided by the provider and are not obtained via a third-party service like Skype. Simple convenience can be a big factor in ensuring the provider gets the revenues instead of suffering an even greater fall in income.

Skype is a particular case in point because it encrypts and masks its use, making it hard to detect and prevent. Consider the sophistication of technology used to monitor flows of IP data. There are multiple business cases for techniques such as deep packet inspection, so ensure that revenue management projections of the potential impact of inadequate monitoring are added to those business cases for a multipurpose IP monitoring solution.

One paradox of selling "eat-all-you-want" services is that the majority of users will use a modest amount, but a tiny number of mega-volume users may consume all the bandwidth available. As such, CPs need to think ahead and insert some kind of cap to the "eat-all-you-want" service in the contract terms and conditions. Even if this is not habitually relied upon, the communication provider needs to actively monitor for high-volume users and decide, if appropriate, to politely remind them of the cap in their contract. The costs of managing and billing for metered usage may be disproportionate to the benefits; it may be more suitable to use focused customer contacts to manage the relationship with a handful of mega-users who place a disproportionate burden on the network. Indeed, actively managing high-volume users is the precursor to identifying and managing fraudsters and other undesirable users of the network.

Use of a WiMAX network to offer unlicensed telephony services, especially long-distance and international services, is inevitable. "Gray" and fraudulent providers can tap into the kinds of free wireless connection that are typically offered in public places, or they may hack into a paid service. Indeed, they may actually pay for a paid service, if the margins are good enough. Note that sometimes the best customers for one service are actually using it to steal a different revenue stream altogether! Whether the access is free, hacked or paid for, it can then be utilized to undercut the usage-based charges for voice calls, especially high-margin calls to international destinations. Worst of all, if fraud prevention processes are lax, there will be no information to identify who the customer really is and there may be no address for them. As ever, prevention is better than cure, and it is vital to aggressively deal with the businesses that advertise services that can only be offered using frauds or gray traffic. However, if prevention is failing, look for the following signs of trouble:

- High levels of use
- Unusual use patterns such as extremely long periods of use that cross between day, evening, and night
- Varied traffic destinations
- Multiple simultaneous events

When detected, the fraud manager needs to disconnect or block the user with extreme prejudice. Block first and ask questions later. However, if the service is offered in a public place, it may be worth muting the initial block by stepping down to a more specific filter of certain kinds of traffic like VoIP. This then permits legitimate uses (people checking emails) while stopping the revenue-eaters who will be causing congestion for ordinary users anyway. Standard network infrastructure may not support selective blocking, again giving rise to a business case for increased revenue protection capability, though, as always, a rational judgment of costs versus benefits should follow from meaningful and real data, not from guesswork. It is the job of RA to do the calculations, and even if the costs are too high relative to the benefits at first sight, RA should keep on monitoring in case a change in traffic and use patterns shifts the equation in favor of deploying enhanced revenue protection technology.

While WiMAX gives rise to new risks, they are not entirely new risks. The trick is to recast classic techniques of RA and fraud management and redeploy them for a new service. Treat the case on its own merits, but monitor and gather data so the business makes an informed decision about the revenue cannibalization and revenue loss it endures.

Chapter 11

The Business of Revenue Assurance

Revenue Assurance Caterpillar

Eric Priezkalns

Caterpillars are ugly. They crawl around and eat a lot. Butterflies are beautiful. They fly. Is revenue assurance (RA) in the process of transforming from a caterpillar into a butterfly? Vendors that used to shout about RA now speak the words in hushed tones, instead talking about the next big thing, which may be business data analytics, revenue intelligence, business assurance, or pretty much any combination of words that implies "better and more than current revenue assurance."

Of course, transformation may not be good for everyone. It can open the door to new entrants in the market, and this would offset the consolidation amongst RA vendors that has been seen in recent years. Business intelligence and other IT firms might start seeing potential in entering the RA domain. Even if the RA market is small, larger players may enter as a defensive response to the threat posed by the expansion of niche RA vendors. This has the potential to create a blurring of the distinction between what is an RA system and a system that happens to be used for RA but could also be used to meet other objectives as well.

The rest of this chapter talks about trends in how the market for RA software and equipment has changed and is changing. Change is good, but not for everyone. The chapter and this very short article might make uncomfortable reading for any vendors wanting to hold on to old-fashioned ideas about the role of RA within a modern communication provider (CP), or who are unable to invest in the research and development needed to expand beyond the RA niche. But if you want to be a butterfly, you have to say goodbye to the caterpillar.

Horizontal and Vertical Expansion of Revenue Assurance

Eric Priezkalns

The business world is full of horizontals and verticals. If I can sell a chocolate desert to your mother, maybe I should try to sell a chocolate cookie to your daughter. That is horizontal thinking. If I can sell the chocolate desert to your mother, maybe I should buy the business that sells me the cocoa beans. That is vertical thinking. Horizontal and vertical thought processes apply equally well to every enterprise currently working in the realm of communication provider (CP) revenue assurance (RA). Because of changes taking place in the market, this is a good time to pause and apply that analysis to where RA is today.

In recent years, we have seen the "core" RA market nearing saturation. First-time sales are dwindling and the annual revenues from upgrades and maintenance will not fully compensate. The sustainable market for RA in CPs is less than the peak market of the last few years. The RA market enjoyed a gold rush when the number of CPs with RA systems started to outnumber the ones without. This prompted the remaining CPs who were without RA tools to play catch-up and go out and buy COTS solutions, for fear of being at a disadvantage to their competitors. "Me too" became a big driver for sales of RA systems. However, any gold rush must come to an end when the gold is all mined. The inevitable plateauing of sales has driven cost cutting and consolidation between RA vendors. Consolidation will not end until the majority of vendors are profitable, (or, less likely, when they have been consumed by much larger businesses willing to sell RA without turning a profit, just to get access to CP customers). Few enough of the current RA vendors are profitable. Apart from ruthlessly cutting costs and trying every trick to win new work ahead of rivals, while using every incentive to discourage existing customers from swapping to rival systems, what can an RA business do to rejuvenate revenues and deliver increased growth to investors? The obvious solution is that they can expand their business to new markets. The really interesting comparison is the different ways they now pursue expansion.

The vendors in the CP RA niche have benefited from the effect of sharing a common definition of their product. It may sound counterintuitive, but if your rivals sell similar products to you in a growing market, it can help you to sell yours. If five guys try to sell the same thing with the same, common, product definition, then you might buy from the fifth guy even though you first heard about the product from the first guy. There was a time when nobody used the phrase "revenue assurance" or knew what it was. By concentrating their marketing efforts on a common solution set, the vendors created a common brand for RA and helped to define what it is. This enabled the RA market to grow more rapidly than if the vendors sold disparate solution sets with differing names and descriptions. The problem for the vendors is that now that the market is mature, the commonality of the phrase "revenue assurance" has stopped being an advantage and has become a straitjacket.

The phrase "revenue assurance" has been a misnomer for a long time already. In Europe, RA started with ensuring completeness of revenues for retail telecommunications products, and many other parts of the world had the same starting point. When selling products to ordinary people who receive a bill and pay it, or buy a voucher and top up their prepay account, the only concern is to get all the revenues due. As soon as you move sideways into wholesale and interconnect products, then you need to deal with costs as well as revenues. US vendor Connexn used to insist on using the ugly three-letter acronym that was "CRA," which stood for cost and revenue assurance. Thankfully, Azure killed all talk of CRA when they bought Connexn out, but the existence of the phrase highlighted the relatively high importance attached to managing costs in the North American market. The meaning of RA in Europe has been steadily stretched in one direction, but North American CPs came at the challenge from a different direction. European RA has generally extended its scope from revenues to costs; North America has moved from cost management toward assurance of revenues. As a result, the North American market continues to reserve the phrase "revenue assurance" for activities that deal with revenues alone, and do not blur the edges of the scope of RA with that of cost management. Meanwhile, in regions like South America, RA has been instigated in an environment where wholesale and interconnect were given at least equal priority to retail, making both costs and revenues equally central to the mission of RA from day one.

International vendors attempt to sell in all markets, and understanding nuances in what people expect from RA makes a difference to how successful they are. Many a vendor has lost a sale because they failed to find the right pitch for a product that was actually competitive but not properly understood by customers. Salesmanship involves knowing the personal buttons to press to motivate individuals, but it also involves finding the right language to express a proposition that will resonate with the potential customer. The technology for RA is far from unique. Its distinguishing characteristic is that it is packaged as a solution for a certain kind of problem. The ability to diagnose, define, and talk about the customer's problem, and adapt the technology accordingly, makes a great difference to the vendor's chances of making a sale.

Revenue assurance for CPs is a well-defined solution space within the universe of data quality. The challenge for vendors in a market that has stopped growing is to lever themselves into new solution spaces without abandoning all of the advantages of the generic brand of RA. A business that sells technology to interrogate and improve data—in essence, what all the RA vendors do—can find lots of possible applications. The trick is to create a compelling proposition on the basis of using the technology, while understanding the customer's specific problems and giving them an efficient answer without reinventing the wheel each time. All vendors are pondering how to rebadge their current products to enter, or sometimes make, new markets.

In the last couple of years, vendors have started making interesting decisions about how to move vertically or horizontally to offset the slow down of sales in the core RA market. A vertical expansion of offerings means selling more types of solution to the same type of customer, while a horizontal expansion of offerings means selling the same type of solution to more types of customer. A simple comparison should illustrate. When Subex bought Syndesis, they leapt into a new market: service activation. As a consequence, it is now appropriate to describe Subex as an OSS vendor, although they started out by selling solutions that fell under the BSS banner. Expanding into service activation was a vertical expansion, because it meant Subex could sell a greater range of solutions to the same customers, namely CPs, and because the technology has interlinks that create new assurance possibilities. In contrast, WeDo have proudly reported their growth in revenues from other businesses than CPs. As often noted on *talkRA*, there are analogies between RA in communications and sectors like transport or banking. WeDo has expanded horizontally, by repositioning its technology to provide solutions to other industries, moving outside of the telecoms vertical.

In many ways it can be easier for a smaller business to expand vertically rather than horizontally. A good business builds up a reputation in one sector, but is unknown in others. If you expand your offerings vertically, you can sell more to your existing customers. If a customer likes a current vendor, they may be the one to instigate the vendor's expansion by asking for new solutions from the vendor they already trust. When expanding horizontally, it is difficult to translate the reputation in one industry to credibility in another, and it is challenging to make the new contacts needed. There are good reasons why WeDo is unusual in successfully expanding horizontally: they utilize home field advantage by concentrating on Portugal and levering common contacts, language, and culture.

For most players in the RA sphere, vertical expansion has been the order of business. A vendor that only offered switch-to-bill would also develop a network-to-bill product. A vendor that focused on FMS would expand into RA, and vice versa. Retail RA offerings have been joined by solutions better geared to wholesale and interconnect. Instead of just analyzing what went wrong, analytics can be used to highlight weak margins or model what might happen if prices are changed. As time passes, there are fewer and fewer safe ways to expand vertically, and vendors are taking different strategies. The riskiest was Subex's move into service activation, but there are other examples. cVidya has repositioned itself as covering dealer management in addition to RA, and has finally started to offer FMS. More and more vendors are playing down the niche that helped them for so long—RA—and are devising new names to cover their expanded field of interest.

Increasingly we hear more and more talk of "business assurance" or "profit max-imization." The idea is to create a new brand that encompasses the old brand of RA, while allowing the vendor the opportunity to move into areas of potentially higher growth.

One problem for vendors is that the new brands are fragmented, losing much of the value that came from the commonality of RA. If only one vendor uses a phrase, it has no recognition value beyond that vendor and few outside the vendor will repeat it. Instead of the unconscious team effort that grew the RA market, we see disparate marketing pulling customers in different directions, and hence leaving them less convinced than if every vendor pulled the same way. At present, there is no dominant market leader in RA, or we would see a pattern of one business decid-ing to expand in a specific direction and others following it. We nearly saw Subex attain the necessary dominance. Some other companies have cautiously echoed Subex's talk of a "Revenue Operations Centre," but Subex never gained critical mass and many rivals simply went their own way. The result is a fragmentation of RA, with disparate interests taking the idea in different directions. As time passes, the fragmentation will get worse. It will only stop if one player adopts the mantle of consolidator and can convincingly position itself as both a genuine thought leader and the market leader. If this happens, they will be able to influence the thinking of a large enough share of customers and hence set the mould for rival vendors to copy.

A final observation is that the RA space does not exist in a vacuum. Even if you have no knowledge of the rest of the world, it still exists. While RA vendors want to expand out of the CP RA niche, other businesses want to expand into it. Because RA is a subset of the data quality world, there are plenty of businesses that could move into RA on the basis of their existing skills and reputation at handling data. We see this already with some of the alliances announced between RA firms and data appliance and warehouse businesses. The RA firm brings vertical expertise and their partners augment the technology thanks to the strengths they generate from horizontal scale. Some of the RA vendors must be hoping to find a buyer that wants to move into CP RA without going to the trouble to do it organically. In contrast, the current players in the RA space will fear that bigger businesses may try to grow organically into the RA market, causing even tougher competition. Companies like ACL have made organic inroads into CP RA and the risk remains that bigger businesses will do the same.

When a market is growing, it is easy to get consensus on what to do. Look to see if the next guy is successful, and if he is, then do what he did. That is why we had so many vendors describing themselves as selling RA. It also explains why, at the tail end of the gold rush, individuals working in CPs now want to sign-up and be certified to work in the field. Meanwhile, the world keeps on turning. Now the RA market has reached the top and will settle for smaller ongoing revenues, it gets harder to decide what to do next. Success is less easy. At the same time, you have to make decisions because you cannot afford to stand still while rivals innovate and experiment by moving into new fields. For a long time, I tried to tell people what RA is. Now that they know, it is nearing the end of the life of RA, or at least the end of this phase of its life, as it prepares to transform like a caterpillar turns into a butterfly, or to give birth to new business propositions. Vendors need to make tough decisions about their business direction and future. We will not know which decisions were important, until we can look back on them and see what the results were. What we do know is that now is the time for making those important decisions.

Spicing up the Market

Eric Priezkalns

When you hear people talk about possible futures for business sector that sells revenue assurance (RA) to communication providers (CPs), you tend to get one of the few answers:

1. It becomes a well-defined niche in the ever more sophisticated world of CP BSS software. The job of people in CPs ends up being to run that software in the equivalent of a revenue operations centre.
2. It gets embedded, SOX-style, into corporate governance frameworks as a way for big business to both execute and demonstrate management control over its underlying data. The main output is a tick list to show the auditors and ultimately the shareholders that everything works correctly.
3. The focus remains on generating quick-win extra money for businesses, so demand for RA ebbs and flows with the perceived need.
4. It gets absorbed into business intelligence, with assurance being one of many by-products of having a complete view of all the company's data.
5. Other industries emulate the approach of CPs and do similar things. This creates new opportunities for the transfer of the RA skills and technology from CPs to other industries. Alternatively, CPs themselves gradually adopt new business models and RA migrates into new business challenges, while remaining within the same company.

Any or all of these may be true, and I would not rule out any possible combination of the above. What I have often said is that there is no great future in a view of RA that remains fixated on the nuts and bolts of current CP systems and processes. Many may understand RA purely in terms of the detailed workings, but if you walked into a bank and tried to persuade them of the relevance of RA in terms of switch-to-bill CDR reconciliations you would expect plenty of blank looks. Similarly, if a CP's business model changes, then its RA function will end up in a smaller and smaller niche unless it learns the happy knack of applying its framework of goals and techniques more widely.

RA can be applied more widely. Much of what is tightly branded as "revenue assurance" in CPs already exists in other businesses: they just do not have a consistent name for it. Banks do the equivalent of RA, but they would not call it RA because it is so embedded into their way of working that there is no need to recognize it as a separate discipline. Think of it this way: if your retail bank was as error prone as telcos reportedly are, you would be carefully checking every line of your statement each month. Similarly, and more obviously, RA also exists in the utilities sector. A lot of readers will have heard of the publicly listed British company called *Revenue Assurance Services* (RAS). It focused on the utilities market, predominantly gas. Their remit was the abstract equivalent of RA in CPs. RAS offered outsourcing and consulting services in the key areas of debt management, meter readings, bill estimation, reconciliation of provisioning to billing, and bill verification. Some of this has a straightforward analogy to the kind of RA found in CPs. Recalculating bills to ensure the billing system logic and reference data is correct would be the same for either sector. Other activities have little or no analogy. CPs need not worry about regularity or adequacy of customer site visits to read physical meters. However, one area sticks out as being common to CPs as well, but is often left outside of the scope of CP RA: collection and management of debts, the "fourth C." I think this is because debt collection is not a challenge specific to CPs. It is an area where you can reasonably recruit experienced staff from other industries.

Most (but not all) viewpoints on RA in CPs associate it with problems unique to CPs. On one level, this is an advantage. If those problems are unique and significant, then the scale of error may be large and go unnoticed for a long time because people lack the experience and conceptual models to identify the risk. A narrow focus will give the best results. On another level, this is a disadvantage. Setting the scope of RA according to a narrow view of existing CP systems, processes, and issues makes it harder to transfer experience to other business models and increases the chance that the CP version of RA may become outmoded and redundant.

I talk about Revenue Assurance Services in the past tense because it was taken over by Spice, the support services company. The acquisition is confirmation of the perceived ongoing profitability of a business that aimed to increasing the returns of utility companies, across a broad variety of technical and process areas. It is also a warning shot that the Revenue Assurance Services business model could be successfully extended beyond utilities to other sectors, including communications. While the first goal will be to grow sales to electricity suppliers, Spice's reach into telecoms, retail, and banking must mean they will be looking at any opportunities to extend their own RA formula further. Areas that are easy to outsource or which are relatively similar across sectors, like debt management or bill verification, will be easiest entry point for a utility-oriented vendor to expand into new industries. If suppliers in the field of CP RA do not sufficiently expand their own horizons, and aggressively pursue growth outside of electronic communications, they may find that the suppliers to utilities have moved first, and are invading their territory.

Listening to a Market Leader

Eric Priezkalns

I first met Subash Menon in Beijing, at an event where we both were speaking. Menon is a self-made superstar of the revenue assurance (RA) industry, building up Subex from scratch, and turning it into one of the best-known publicly traded Indian software companies. While I have not been back to Beijing since, I am sure Subash has, given his international deal-making agenda. Over the years, he has expressed two sets of views that confirm his thoughts on how the revenue management industry will evolve. I want to revisit those views here, not least because I think he is right on both counts.

1. *Fewer people will write code for revenue management software in the future.* Lots of code hackers have done well out of relatively small projects to deploy fraud management and RA software. They may have been employed by a communication provider (CP) to rig some bespoke reconciliation, or they may have worked in one of the many tiny vendors who each sold their own vaporware to the few customers where they enjoyed personal contacts. Often the latter grew out of the former. However, economies of scale means the party is coming to an end for these guys. You cannot effectively compete with an army of developers in low-cost, high-skill countries such as India, especially when their code is sold to hundreds of customers worldwide. Small vendors and in-house developers will, most of the time, and especially in the long run, be overtaken by firms like Subex. Economies of scale, a low-cost base, and educated employees are vital to delivering RA products and services that are of superior higher quality yet are offered at a keen price. The small and in-house players may get specific results quickly, as a consequence of their adaptability, but as they produce something that suits one or perhaps two customers, they do not build an offering that will be just as relevant and marketable to other potential customers. That means they have to milk the customer for maintenance fees, a situation that can last a surprisingly long time, but cannot last forever. This also means that those people who tried to have a RA career by trying to combine being an okay hacker with being an okay business consultant need to stop being okay at two things and start being good at one thing (at least).

2. *In the future, traditional revenue assurance will be a part of service assurance.* Some traditional views on RA treat it as a subset of the responsibilities of the Billing department, or think RA is about finance, creating and running their own automated checks every day. That way of thinking belongs in the past. That kind of RA is all about data, some of which is about billing, some of which is about the network, some of which is about other things such

as the data sent between businesses. There is plenty of data in billing and customer relationship management (CRM) systems, but the people with jobs that require access in real time to data of the highest integrity, from the source systems, and in the most detail, are the people working in service assurance. Compared to the problem of managing data for service assurance, RA is easy. It is much easier to extend service assurance into RA than the other way around. So instead of people looking at problems in billing and then trying to work out how that relates to a network they are ill-equipped to understand, you will have people who understand the network, and have all the data about it, steadily expanding their influence and control of BSS systems. CPs will have to do this with their network-facing teams, as customer charging is increasingly going to be performed in real time anyway. It will be too expensive to keep trying to train finance and billing people to (poorly) understand a network when you can more easily train people who understand the network on the importance of downstream data integrity. So automated checking of downstream data is going to increasingly be part of the responsibility associated with service assurance. If RA persists as a separate activity, it will be because it is about the things that cannot be done by simple brute force using existing data: analyzing future risks, defining controls to address those risks, setting objectives for the data crunchers, and prioritizing on the basis of understanding the financials and wider business objectives.

When it comes to software, big is beautiful. One big development team, whose output is sold to many customers worldwide, will inevitably outstrip the cleverest of small development teams that have only a limited number of customers. Link RA to service assurance and you extend development and increase synergies even further. Small players need to find new niches or can expect to be consumed by the bigger beasts of RA. Menon has proven himself to be one of the great survivors of the industry. The world is changing; which of his rivals also have the instincts needed to survive?

Chapter 12

Power to the People of Revenue Assurance

Common Theme to Good Training

Eric Priezkalns

In a perfect world, everybody would get one-to-one tuition that was tailored for their every need, whenever they received any kind of education. In this imperfect world, our kids go to classes where one teacher takes a lesson for many children, students may fill large halls to listen to the same lecturer, and lots of different people will read the same book to get the knowledge they need. In this imperfect world, what is common to good revenue assurance (RA) training, and what should every RA practitioner learn?

A regular complaint made by pupils about the RA training they receive is that it fails to be specific. Though the complaint is understandable, it is also indulgent of the perspective of the pupil. Everybody wants to know what he or she needs to know. But who knows what you need to know, better than you? And why should the RA practitioner in one communication provider need to know the same detail as a practitioner in another communication provider, when those two providers use different BSS to charge for different products sold over different kinds of network? Splitting the world between fixed-line and wireless providers hardly helps; from an RA perspective, a retail-oriented fixed-line communication service provider (CSP) might be more similar to a retail-oriented mobile provider than it is to a wholesale-oriented fixed-line CSP. The complaint that training is not specific enough reflects a failure to appreciate a sense of scale. If a pupil of RA finds their training is not specific enough, it is because the work they need to do is specific, and there are too few people needing to know the same things to make it worth trying to get them all into the same classroom.

Indeed, RA training is often very specific, sometimes to the point of not being relevant to the student. The more serious problem with most RA training is that it is overly specific. It can be specific to the point of irrelevance, when it starts to include technical detail of no relevance to the communication provider (CP) where the pupil works. When RA training is over specific, and lacking any common and universal principles, it does not train practitioners to be versatile and to cope with the unfamiliar. Because RA is about dealing with problems that people did not even realize were there to begin with, we cannot train people to be good practitioners by only giving them skills relevant to a few problems we now know to anticipate. The good RA practitioner must be able to adapt their skills to the unanticipated too, and must learn how to find issues even when nobody else has considered they might occur.

The specific instances of RA must have something in common, but identifying what is common is far from straightforward. Shakespeare makes a similar point about dogs in *Macbeth*, because dogs can be very different to each other but still have something in common:

> … hounds and greyhounds, mongrels, spaniels, curs,
> Shoughs, water-rugs, and demi-wolves are clept
> All by the name of dogs.

A good RA training will be based on what is common to every situation. This includes the situations that the student is not familiar with, and even includes the situations that nobody has experienced yet because these have not happened yet. Tailoring a course for your organization sounds efficient, but how useful is the course if your organization is changing? How long will its value persist? Even if tailored, the course must be based on principles that can be applied outside of the specific examples that are covered.

Universal principles, once understood, are more valuable to the student than lots of specific parcels of disconnected knowledge. The clearest failing in the RA world is that lots of people know lots of things about lots of particular detail, but struggle with what is RA in general. Take them from their comfort zone, and they fail. Present them with a new problem that requires skills they lack, and they run from it and search for an old problem they have solved many times before. This observation was one of the driving motivations for setting up *talkRA*—to force people out of their narrow silos, get them talking to each other, and make them realize that RA is bigger than the skill set and experiences of any individual person. It was also motivation for this book, which stubbornly explores what is common to all RA and avoids putting RA in a comforting, but constricting, straitjacket.

At its core, any training should be based on universals. Practitioners have more valuable skills if they learn methods and techniques that can be applied to any situation, instead of learning how to do just one task. It is the same as the differences in how we might teach history. I can teach somebody history by making them memorize a list of dates and events. I can teach them a lot more history without mentioning a single date or event, if I teach them how to do their own research. If students can do their own research, they can then find out the details that they need, when they need it. The most valuable kind of training ensures that universal principles are explained, and then made specific and relevant to the audience, depending on what kind of audience is receiving the training.

Unless specific training is consistent with universal principles, two specific training courses simply do not teach the same thing. If you wrote a RA training course for one communication provider based on one set of principles, and wrote a second RA training course for another communication provider based on another set of principles, then you have no consistency in what you are saying RA is. In short, you trained the two classes to do different things, with nothing in common. They may both be good courses, but they cannot both be good RA courses. The better the underlying principles used to create a course, the more universal the principles those are, the better the training is for the recipient. Why? Because the students will be able to reapply those principles to new situations, if their business changes or if they move to do the same job in another business. Otherwise, they will just need to be taught everything afresh every time their situation changes.

You can teach people to do a job a certain way, by training them which buttons to push and how to use some software. They can do that job perfectly well if they keep pushing the same buttons, even if they have no idea why they are doing it. Then swap them over to new software, a new company, or a new product to be assured. You have to train them to push new buttons, and the training begins right back as if they learned nothing before! It is better that they understand what is common between the two scenarios. It is not just about being efficient with training, it is about developing people as people—encouraging them to think and be adaptable, teaching them principles they can observe and reapply. Nobody consciously decides to be a mindless drone who needs to be reprogrammed for every new task that is provided. Of course, cynics can make more money by exploiting mindless drones: the drones will be made to pay over and over again for more and more training.

There is lots of bad training in RA, and we need to identify why. There are lots of people, with very limited experience, offering to teach people who work in situations that are very different from any they understand or are familiar with. There are also lots of people happy to be trained in a kind of RA where they just want to be told how to push the buttons, and not to think for themselves. Those people might do okay in their job, but they do not understand RA and will be little better than a complete novice when they change jobs. Worst of all, this sector is full of people who know how to do one thing and then pretend that one thing is the same as everything in the world of RA. They claim their one technique is as equally powerful, relevant, and applicable to every business and every challenge. They train other people to do that one thing, fooling them into thinking they now understand everything in RA as well. The good RA practitioner knows many techniques, knows how to use them, knows how to adapt them to be used in new and unfamiliar situations, and even knows how to make and adapt new techniques to fit the task. You can only teach that from first principles and by encouraging people to exercise their skills in the real world, not in the classroom. Good RA practitioners also know their limitations and when they need to draw upon the complementary skills of others.

First and foremost in RA training, people need to understand why they are doing what they are doing. Then they need to understand the choices they have about how they do it, so they can pick the best technique for the job. At heart, giving pupils the ability to chose from a wide-ranging palette of RA techniques is what all good RA training has in common.

Training for Revenue Assurance

Mike Willett

Much is often made of the need and benefit of training to enhance the knowledge and skill of revenue assurance (RA) practitioners. The claims made of the benefits certainly vary considerably and, in recent times, there also seems to be an increased focus on accreditation or certification to prove that the student has the necessary skills. This article is not about the content of varying and competing courses but some considerations that I recommend people think about when evaluating different training options.

First, technical training on how to use RA software tools is essential. There is little value in spending money on a tool that no one can use or, as is often the case, is not used to its full capability. There is no need to write anything more on that. We all know that it is people who use tools. They need to apply their thought processes on the best way to achieve the outcomes that are expected of them.

Training in RA often seems centered on ensuring that the students have an adequate understanding of how a communication provider is "put together." By this, I mean ensuring that the underlying technology and platforms are understood. This is certainly valuable but I suggest any RA training you undertake needs to go beyond this and if it is too weighted in this area, then the course is more telco 101 than RA training.

RA training often discusses known leakage points and what to look for. It is great to hear that operator ABC lost millions of dollars because of a wrong configuration, but there is a reasonable chance that the configuration is specific to that operator and not so relevant to yours. Case studies are always of interest but for anyone other than absolute RA novices, we know how leakage can occur. Simply, we lose call/event records, we misalign services provisioned to those billed, and/or we charge the wrong amount. Most leakages seem to be some variation of these. What you want from training is to improve your efficiency and effectiveness at finding revenue leakage.

So how can this be done? I do not have all the answers, but here are some thoughts:

■ Ensure that the training encourages you to think about your own operational situation, as opposed to a generic model of a typical communication provider. For example, event records can be lost but what are the relevant systems in your organization, what are they meant to do, and what might go wrong? With that knowledge, you already have the start of a scoping document for some work.

■ Model what good RA work looks like—how long should it take, how will precision be ensured, how is the program managed, how are updates communicated, how are outputs prepared, what will be done to fix any identified leakages, and so on. This includes understanding what the challenges are to undertaking RA work and highlighting, very specifically, how these can be overcome.

■ Work on defining a program to prioritize your efforts. We all know our companies are large, and we could look everywhere for leakages, but depending on our priorities, where should we invest our time? For example, if you are chasing revenue, then, in general, complexity and/or manual processes lead to the largest losses. Training should ensure you understand some of these principles and apply them to your situation—what is complexity to you (is it in the design, the build, in the implementation, or in what customer service representatives are meant to do)?

To summarize, when you look at training options, be wary of training purporting to be RA when it may be on other subjects and seek out training that defines better how you can specifically improve the quality and quantity of your RA output.

Price of Professionalism

Eric Priezkalns

It is December. Along with the Christmas cards I have received a most unwelcome delivery that I always receive at this time of year: the reminder that my annual fees are due from the Institute of Chartered Accountants in England and Wales. I must pay nearly 300 British pounds (around 500 US dollars) for the privilege of calling myself a chartered accountant for another year. This is not a small amount of money. However, another point is worth making: you get what you pay for. The more value that a profession, membership of an institute, and a qualification generates for its members, the greater the sustainable level of fees. New members will be attracted because they will conclude the cost of entry is worthwhile. People will remain members because they find in actual practice the cost of membership is worthwhile. Membership costs include monetary and nonmonetary elements. Fees and the cost of ongoing education are monetary, the time spent in continuous professional education (CPE) is nonmonetary. At the same time, to buy influence, expertise, and a top-notch educational and qualification program, which ensures not everybody who wants to pass is allowed to pass, costs money. There is a virtuous circle here: value for members gets recycled into fees for the professional bodies, which gets recycled into value for the members. What then is the price people are willing to pay to be a RA professional? Having established the link between cost and value, we can ask the same question another way: what would be the value of being a recognized RA professional?

This is a question about economics. It is a matter of supply and demand. There is demand for RA staff, but can you alter the supply by creating barriers to entry and increasing the value of those who have achieved a certain level of measurable competence? If everybody who wants a qualification also gets that qualification, then the value of the qualification will tend toward zero. If the qualification is not a reliable indicator that the holders will be better at their job than people without the qualification, the value will again tend toward zero. If the qualification is too hard to get, then it will not be recognized and too little investment will be made in sustaining it, which will undermine its value. Finally, if people do not intend to use the qualification to get value, then it will not have value in practice. This may occur if nobody, in real life, ever gets a better job because of the qualification added to their CV—indeed, a qualification that is not respected may be more of a hindrance than a help! Another way that RA qualifications may lack real economic value is if qualified people do not really intend to pursue a career in RA, but instead expect to get promoted up the ranks of their company into other jobs. If they progress in their career by learning how to perform roles where RA qualifications are irrelevant, then the value of the RA education they receive approximates to zero, at least from the perspective of their employer.

I have trained a few people in my time, and my honest opinion is that a minority of those people were seriously thinking of long-term careers in RA. If they were asked to spend thousands of pounds educating themselves in RA, they would not pay it. In those circumstances, it follows that, even during the season of goodwill to all, there will be no fairy godmother, Santa Claus, benevolent employer, or anyone else who would be prepared to pay the cost for RA training that does not add real value to the individual. Investments in education follow the same cost-benefit dynamics as other investment decisions. Whoever pays for the education expects to receive a real return on it. Individuals will pay for it themselves if they expect to get a better-paid job afterward. An employer will train an employee, but only if they find there is a trade-off in having a more productive employee afterward. If the employee leaves, the employer will likely seek to recoup their outlay. Conversely, paying for a long education process may be a way to guarantee employee loyalty without needing the same size of pay packet. Whoever eventually decides to invest in education, the decision is not short term, because no education will generate a positive return in a short space of time. Only people who think sufficiently long term could justify a significant investment in education.

Of the people I trained, all of them were pretty much glad to get the training. Training is one of those things people like, especially if they do not pay for it, and better still, if their employer pays for their time spent in training. This means the value of training cannot be measured from the feedback of participants. Or rather, you get more meaningful feedback from people who pay for training out of their own pocket or who give up their own free time to receive it. This means most feedback is not a good measure of whether a course meets a particular customer's needs, especially if we think of the customer not as the pupil in the classroom, but as the business that employs the pupil. It is also important to question how satisfied people would be if they were forced to take an examination at the end of training and a significant proportion of those people were to fail the exam. The difference between offering training and offering a qualification is the drawing of a line commensurate to the quality of the qualification. Too lax, and the qualification is worth little; too strict, and people will not seek to attain it.

From my own experience of giving training, the key factors that determine the true value of education (as opposed to the factors that are most likely to lead to positive feedback from training course participants) are

■ Being specific to the customer's needs (including the employer's needs)
■ Providing education and, in particular, being able to answer spontaneous questions by drawing on the genuine real-world experience of the tutor
■ Keeping the education cost effective

Specificity is vital to effective education. Telling people things they find interesting, stimulating, or thought provoking is not the same as educating people in things they need to know or stretching their abilities. A junior analyst may gladly sit in a training workshop listening to a discussion of how to influence CEOs or making pricing decisions, but such training is irrelevant to his or her job. However, specificity of RA training is a problem. No two communication providers are exactly alike, and the assurance challenges will differ accordingly. Similarly, technical specifics are vital for some forms of training, but not others. In abstract, the principles used to assure a retail wireless voice usage product are no different to those relevant to a wholesale leased line product. However, the technical details of the products are very different. Managers may need to learn abstract principles, especially if they work for providers that work across many segments. More junior staff need details, but not abstraction. Some of the technological details may be specific to the business, for example, if systems were developed in-house, which puts a very real limit to the value that can be gained from external training in RA. There are also different techniques for how to perform assurance, so there is also the need to tailor the skills taught depending on the techniques the staff are expected to utilize. Some RA experts might like to reduce everything to mining data from a warehouse, which in turn takes feeds from all the source systems. There is nothing wrong with that technique in itself, but there are other techniques that belong under the banner of RA and that may work better in certain circumstances. Other techniques that might be covered in education include risk appraisal, statistics, workflow mapping, or soft skills such as interviewing for knowledge elicitation.

Not all RA employees need to know every technique, but a good RA team can call upon staff who can deploy a variety of techniques between them. In some businesses the RA department is exclusively engaged in automated data analysis or exclusively in process review. Though I might think the best RA departments combine the two, it is not my place to redefine job roles, so cost-effective training for a more constrained RA function will remain focused on skills that fall within those constraints. A more expansive course would cover many of the techniques that would be relevant to an RA department in another business, creating a conflict between the needs of the employee (if they want to move to another job in another business) and the needs of the employer.

There are many suppliers that offer standardized introductory training for RA. These courses are the most generic courses in nature, because focusing on basics avoids the need to deal with the sheer variety of issues that different communication providers, with different products, processes, and systems, will face. Introductory courses are relevant to the largest cross-section of RA employees, not least because a large number of novices continue to be recruited to RA each year. Because the courses are generic, they can be repeated over and over and hence generate the best returns for the effort spent in developing the course. There are far fewer suppliers who offer tailored training or more advanced courses. In a predetermined course, the question is how well the standard course will correspond to the customer's specific needs. It may be better to get tailor-made training, if tailor-made training is available. In any subject, somebody has to go first; just because it would be nice to be trained by someone else does not mean that there is someone else who is competent to give the training and willing to do so for the money on offer.

Genuine real-world experience is a major problem for education. It is hard to learn from a teacher who does not understand the subject he or she teaches. Many people have practical experience of some elements of RA. Fewer people have practical experience of most elements, and nobody has personal hands-on experience of doing them all. Even the most knowledgeable person does not have current hands-on experience of applying all the kinds of skills relevant to RA to all the kinds of communication services offered by all the kinds of communication providers. Each year the business models of communication providers diversify further, increasing complexity and increasing the gap between the most an individual could have learned from personal experience and the extent to which RA has and should be applied across all communication providers. With no genuine institute that can build on the differing expertise of a mixture of experts, the authors of training courses exhibit their personal prejudices to the detriment of highlighting alternative ways to do RA.

At best, the cumulative hands-on experience of the tutor, combined with the tutor's experience of managing other people's work, will cover the most commonly used RA skills and the most common communication provider services, though keeping that knowledge up to date with the latest technology will always be a serious challenge. So one question the customers of education need to ask themselves is how deep and current the knowledge of the tutor needs to be, especially for any specific skills or technologies they need covered. I have worked with consulting businesses that, quite sensibly, have offered several individuals, each with different skill sets, to run distinct modules within a single training course. Training may not be the core skill of the individual experts, but to maintain the quality, the training course may be compiled, overseen, and guided by an expert in training (as opposed to an expert in RA). This way there is always a person on hand who can answer questions on the basis of genuine current experience, while the course itself is designed to meet genuine training needs. The number of people needed to cover the different areas of expertise will depend on how broad and varied the customer requirements are. Of course, the obvious downside is that training from several experts will be more expensive than that from a lone tutor.

In an ideal world, people with extensive hands-on experience in many areas of RA will deliver inexpensive courses widely. The problem, as you will have appreciated, is cost. Hands-on experience does not come cheap. A tutor should reasonably expect to demand a higher price if his or her experience is broader or deeper than another tutor's, so it should follow that the tutor with experience broader and deeper than anyone else's would be most expensive of all. However, someone with those talents presumably could make more money by doing RA than by giving RA training! The tutor also needs to actually do RA to keep his or her skills and knowledge current, as there is no academic infrastructure to percolate the experience of others into education. So I do not believe any individual can claim to be an omni-skilled expert with genuine deep knowledge and current experience across all aspects of RA. The field of RA is not medicine or law or accountancy. Just because RA is broad and complicated and staff need training does not mean that there is enough money in the discipline to justify a training company making the investment needed to develop a good training course. To develop a good course requires an outlay on creating materials and employing the staff required to provide a genuinely high-quality training program across the full spread of RA techniques as applicable to all products. However, even if RA were medicine or law or accountancy, it would still seem strange for a lone individual to claim to be able to give comprehensive training across all aspects. There is a need to balance the extent to which any paid education course is tailored to the customer's requirements and the extent to which the tutor has both domain knowledge and the skills to provide that course, with the amount of money the customer is willing to pay. There is no perfect answer to that compromise; but there will always be a compromise between specificity, tutor experience, and cost.

There is an observation I would like to make about training, based on my experiences as a manager in several big businesses. People are generally very positive to the idea of receiving training, but they may not be clear as to what they want to gain from training. Some people want to do their job better, some want career advancement within the business, others want career advancement elsewhere. A qualification with recognition value is vital to improving career prospects, but it may not be useful in terms of on-the-job performance. In contrast, training that is very useful for improving job performance may be of no aid in advancing the trainee's career. I doubt many people who currently receive RA training have a realistic aspiration to get a better RA job elsewhere. There are not that many communication providers, and not that many jobs working in RA. Worse still, there are relatively few places where a large number of communication providers are geographically colocated, suggesting a career in RA is suitable only for someone willing to travel extensively and relocate from country to country. In practice, many people take RA jobs as an entry into a nearby business, not because they are dedicated to RA as a discipline or have the intention to move to another home later in life. Career advancement within a business should be based on the employee's results, and not on the basis of some relatively trivial training course. My training as an accountant was long and intensive. Many of my peers failed to complete it. It was not a few days in a classroom after which everybody received a certificate. Employees need to be clear as to how significantly their skills and abilities really are enhanced, in terms of both perceived and actual value, as a result of any educational activity they engage in. That leaves improving job performance as the final reason for training. How much job performance is improved will be entirely determined by how specific the course is, and how much relevant practical knowledge the tutor imparts. For senior RA staff, the training should not be a substitute for clear communications from executives as to what they expect and want from RA. For junior staff, training should not be a substitute for guidance from their line managers.

As a manager in RA, I often put more emphasis on informal kinds of learning than classroom-based teaching. By informal learning I mean activities where staff learn from each other and are encouraged to search for and find answers from all sorts of resources, whether described as "revenue assurance" or not. For example, if somebody needs to learn statistical skills, they can get exposure to materials that will help with statistics; if they want to learn how other people solved a certain kind of RA problem, they can communicate with colleagues inside and outside the business; if they want to know about a product, they get trained by the experts on that product in the communication provider; and if they want to know what their colleagues in the RA department do, they are encouraged to spend a lunchtime listening to a presentation from a colleague, and to give one in turn, and to share information with other group RA staff via the corporate intranet and similar resources. Often an employee's peers and informal contacts can deliver better education than external tutors, and do so at little or no cost. In addition, informal education places the emphasis and responsibility for improvement with the recipient, and hence generates the best returns for the people who actually value education the most.

Informal training is often underexploited by RA teams, even though it has more effective value than formal classroom-based teaching. "Brown bag" knowledge-sharing sessions are a great example of how effective informal training can be. The idea with brown bag sessions is that once every few weeks or once a month, people give up a half-hour of their lunch break to listen to a colleague explain some aspect of his or her work. They can eat lunch (perhaps brought in a brown bag) while listening to the half-hour talk. This is followed by a relaxed question-and-answer session. Because good RA will involve practitioners regularly learning new things—about new products, new technologies, newly identified leaks, new threats to the business—the brown bag session is a way of recycling learning throughout the whole RA team, so everyone understands everyone else's job. Good RA involves good communication skills, so the session also helps to develop the skills of the person who leads it. The presenter benefits from being in a friendly environment where he or she can get feedback on how well he or she presented information and explained complicated topics. If willing, employees from other parts of the business can also be invited to lead a brown bag session, thus broadening the knowledge of the RA team. The brown bag sessions are doubly useful to managers because neither attendance nor giving the presentations need be mandatory—and hence they give a good indication of who in the team wants to learn and share and who only has an interest in education when it involves sitting at the back of a classroom and getting a certificate at the end. When run regularly by an enthusiastic team, these sessions can be very popular, and other parts of the business may increasingly want to join in!

Another powerful yet informal learning technique is the use of a wiki to document RA root cause analysis. A wiki is a web-based content management tool, where users can quickly create new web pages that link to each other. High-quality wiki software can be obtained for free and quickly installed on a web server. A short training session is sufficient to get RA practitioners competent to use the wiki to document the problems they are working on. For RA, the wiki can be treated as a readily visible notepad, where people jot down key facts, theories about what went wrong and task lists for what needs doing next. The advantage of using a wiki is that, being a web-based, visible, and simple tool for documentation, it means anyone participating in the effort to analyze and investigate root causes of a problem can simply look up the relevant intranet page to find the status of investigation work and remind themselves of pertinent facts. This is much better than expecting teams to burrow through their email inbox, looking for messages cc-ed to hundreds of people. For example, instead of writing information in the emails,

RA staff can simply email out a notice that the information has been updated on the wiki. Because the wiki is editable by staff across the business, and it maintains a history of all changes made, anyone who spots a mistake or misunderstanding can correct it immediately. In addition, somebody investigating one aspect of the root cause analysis can use the wiki as a shared, central notice board where they can show everybody else what progress they make, as they make it. The wiki is not restricted to only the RA team; anyone in the business can be granted access, enabling the RA team to pool knowledge and work efficiently with people from all around the business. Finally, because the wiki can be used to document the histories of investigating a leakage from first detection to resolution of the root causes, it grows into a living archive. This archive is useful for subsequent training. When similar problems are found in future, a search of the wiki will highlight how things were tackled previously. Furthermore, as the archive builds, it essentially contains a repository of case studies, highlighting where leakages can occur and how to deal with them. This can be used to train new members of staff. Because the wiki's content is specific to the communication provider, it offers a uniquely focused resource for learning about leakages and controls in the business, but is assembled without the overhead normally associated with writing in-house documentation (which too often ends up dusty on a shelf, anyway). Because the wiki can keep on being updated and changed informally, with minimal fuss, it has a much better chance of remaining relevant. This is crucially supported by it being used in a dual role as both communication aid and information repository.

After discussing the positive elements of informal training, here comes the unpleasant bit about formal training. I do not think that copying other professions has much value to RA. RA will never be a profession like accounting or law, because there is not enough money to be made from doing RA. There will never be enough people doing RA, even when you count everybody who does it on a global basis. Arguably, RA demands a hybrid of many skills, so it would be unrealistic for its practitioners to be as highly skilled in all those facets as is expected of individuals who work in focused professions. Informal education and development is the natural stage of evolution for RA at present, not the codification of a discipline that lacks an academic infrastructure and lacks the economic means to create one. The *talkRA* Web site and this book is our own small way of taking an alternative way of promoting the discussion and education of RA; we must try to practice what we preach.

Finally, you probably appreciate that I have provided RA training from time to time. However, in general I do not regularly pursue it as a source of income. It is just too hard to provide a high-quality course that really meets the needs of individual customers without putting a lot more time and effort into developing a course than can be justified by the kinds of fee people are willing to pay. As in all things, the economics of supply and demand dominate, and what demand there is happens to be insufficient to compete with the other demands for RA practitioners that can, in the end, make more money by doing RA than by teaching RA. If others prefer to teach, that is up to them. If they can earn more by selling qualifications, let them do so. The market will decide what the training and the qualification is worth, not me. My only advice is simple: people should try to ensure they get what they pay for. Expecting more than that is unrealistic. Getting less than you paid for is always a risk.

Standardization Based on Generalization

Güera Romo

While I still have plenty of cleaning up to do of the literature review for my academic research, I am finally arriving at planning the fieldwork of the study, which made me reflect on the discussions I had with a number of people on the standardization of revenue assurance (RA). I do not need to mention the Global RA Professional Association's drive in the certification of RA personnel to point out that there is an obvious movement toward standardization or the implied assumption that, by some miracle, standardization has already occurred.

MTN Group, the telecoms group headquartered in South Africa, is on a drive to standardize technology across its operating units (OUs). According to Deidre Ackermann, group CIO, MTN needs to consider how it implements operating models to attain efficiencies without slowing time to market or reducing innovation. Add to this a healthy dose of diverse cultures and physical locations within MTN's footprint and one can appreciate Dr. Ackermann's challenge. She is under no illusion that using a standard framework such as NGOSS, modeled and implemented, will solve the problem. She acknowledges the fundamental business practices that must change to comply with standardization and that this change is not only affecting technology but business processes as well. MTN's drive is on the total of its systems but let us consider its RA technology across its footprint.

Do we mean standardization of RA tools with the inclusion of all its new requirements or are we addressing the process of RA as well? The literature reviewed for my study provided some insight into the process of RA, which contributes to the list where there is still plenty of clean up to do. There are a number of white papers or magazine articles that propose a basic process of RA execution. A generalization of these sources would conclude that the process steps are Identify, Prioritize, Correct, and Follow up.

However, this is at a level three business process. What happens at levels four to six? Are we still driving standardization if we maintain that standardization goes to operating model level only, or at best the adoption of an industry standard that proposes best practice at level three? Do we adopt Six Sigma ... only to realize Two Sigma because that is all we could afford?

In theory, I agree that standardization will benefit the organization. One can create centers of excellence and I would be very surprised if the RA maturity assessment did not show a marked improvement, but have we considered the cost-benefit ratio? A few thought leaders discussed the concept of RA being part of risk management and several people supported the concept of only doing something when the benefits outweighed the cost. What is the financial and emotional cost of driving such standardization?

Now assume that there is industry agreement that we will drive standardization to its *n*th degree. How do we approach it? Do we select a sample of organizations that we believe are doing RA right, document what they do, and implement that at others? If it is sample based, how big should the sample be and what are the characteristics we should consider for inclusion in such a sample? I have a distinct vision of Noah counting the animals two by two. Can we drive the standardization of RA independently of the standardization of technology and the implied operating models? Can we simply ask the Global RA Professional Association to promptly issue certificates of process standardization being achieved and demonstrated?

Leadership is concerned with the ability to give guidance within the constraints and boundaries of the objectives to be achieved. The "what" is important. "How" is a function of execution against that "what." As long as the business "what" has been achieved, does it really matter how it was done? Unless of course the "what" could not be achieved. If, in the MTN scenario, the drive to have centers of excellence and recognition of skills and people empowerment become the "what" objectives, then the "how" would certainly include something close to a level four or five standardization.

I am not sure we are all clear on the degree of cloning we are proposing.

Which One Are You?

Eric Priezkalns

People who write or talk about revenue assurance (RA) often make sweeping generalizations. So who am I to be any different? But instead of generalizing about the causes of revenue loss or what is the likeliest cause of leakages, I want to generalize about something different. Here is my Chauceresque index of personalities: the kinds of people you find working in RA!

■ The Pioneer. First to identify the need for RA, he has defined the subject area for his business. Used to standing on his own and far from the madding crowd, he forgot how to relate to normal folk after a while. If he says that something is a problem, it is. If he says it is not a problem, it is not a problem. If he is happy with a solution, it is a good one; if not, it is bad. Better not argue with this guy, because he knows you are wrong! He walks a lonely path, wondering why more people do not follow in his footsteps.

- The Hamster. The hamster loves to scamper around in his cage. He finds things, then fixes them. Then he finds more things, that look just like the things he found before, then fixes him. He runs around and around on his metaphorical wheel—a database in real life—for hours, and every hour he spends on the wheel makes him blissfully happy. There is nothing better for the hamster than to feel like he is getting somewhere when he is really not going anywhere at all. Let him out of the cage, though, and he soon gets scared and asks to be put back in.

- The Policeman. This guy is like a rule-bending cop from a 1970s television show. He knows the word on the street and what is going down. His snitches keep him informed of everything that is happening, unlike those pampered executives and the other fools who surround him in the communication provider he works for. The policeman is happy to rough people up and cut corners to put the world to rights, and has no time for people who think he is opinionated. Better not point out all the mistakes he makes, as he can get very angry!

- The Secret Policeman. The secret policeman is quite like the policeman, except a lot more secret. He is so secret that nobody knows how important his work is. Even his bosses cannot be trusted to understand, so he consistently manages communication into bite-size chunks that he feels people can handle. Much of this communication is, well, not exactly true, but the secret policeman cannot afford to tell the truth. Nobody else can handle the truth. The secret policeman thinks everything will fall to pieces without him, which is why he keeps such a close eye on everyone else, without letting anyone keep a close eye on him.

- The Journeyman. Wondering why his career is stagnating, the journeyman started by looking around for a job in his communication provider that he could hold on to despite the constant downsizing. Low and behold, he hears of RA and decides this is his opportunity to give his career a jump start, without needing to really go to a lot of trouble like learning new skills or getting a qualification or any bother like that. Repeating the babble he hears other people say, he has found a career that will keep him feeling important and satisfied for the rest of his life, as everybody tells him he should really be reporting direct to the CEO and he would increase profits by 20% if only people did as he said. As he is a true believer in the religion of RA, there are always new things he feels uniquely equipped to handle that cannot be trusted to anyone else, like SOX or business assurance or risk intelligence or whatever vaguely defined new objective that somebody has described as the next big challenge. RA turns out to be the perfect job for this man, who likes to feel mission critical, but never wants to work late or be disturbed while on holiday.

- The Guru. He talks a lot about RA, writes a lot about RA, and is great at generating theories about RA. The only problem with the guru is that he has not done any actual proper RA work in the last decade (or two).

- The Traveling Salesman. The salesman sells it, but is not foolish enough to buy it. He talks great talk when in the room, but after he makes a sale, you will never see him again. He keeps moving from place to place, worried in case someone catches up with him and wants their money back! Hamsters love salesmen that offer to supply big wheels for hamsters to run around in. Journeymen love salesmen because they think if the company spends a few million dollars on his stuff, it makes the journeyman's job secure. Gurus hate salesmen in general, but if a salesman is willing to give a guru money, or at least free flights and hotels in exchange for the guru's talk, the guru will love the salesman. Indeed, given the right incentive, the guru will probably conclude that whatever the salesman sells is the next big thing in RA.
- The Dead Man Walking. He writes a blog where he makes fun of what he finds in RA. The idea is that if he makes fun of everyone and everything, then it will be impossible to accuse of him of picking on anyone in particular. This would be a great idea except that some people have no sense of humor!

Jobs in Revenue Assurance

Eric Priezkalns

By now you have realized this is not a classified advert, offering exciting new roles in the field of revenue assurance (RA). Do not get me wrong, I think RA is a noble line of work, but I pity anyone looking for a job in RA. Part of the problem is that those jobs are so varied. If you search through RA vacancies, you might find jobs that are labeled "revenue assurance" although they have absolutely nothing in common with your experience of RA. Some RA jobs are about a rarefied accounting discipline where the objective is to ensure that revenues are correctly recognized in the accounts. That is especially common in businesses and groups headquartered in the United States, where correctly following the exact wording of the accounting policy is a very big deal. Other RA jobs are all about niche kinds of IT development. They want people to do wonderful things with code and databases and such, creating new tools and products that, when switched on, will automatically put the world to rights. Now, do not get me wrong. I think that clever revenue recognition accountants and clever IT developers are all smashing lovely people, with important jobs to do and that they add a lot of value to their businesses. But you could hardly think they do the same thing. I would bet good money that there is not a single person on the planet who could honestly claim to have done both jobs. There is no point trying to do it all, because if you try to be a jack-of-all-trades, you end up being master of none.

It gets worse. Most jobs ask for umpteen years of experience working for communication providers. Yeah, right. Because all communication providers are just the same as all other communication providers? They are not, of course. Trust me, everyone thinks like that until he or she tries working for a completely different kind of communication provider. At that point, he or she gets a nasty shock about just how different those businesses are. The similarities between, say, a voice reseller, a wholesale carrier, an ISP, and a wireless virtual network operator are pretty abstract. There is a world of difference between being the kind of communication provider that buys its services from the incumbent and sells it to domestic customers and, say, the kind of telco that is the incumbent. If the species of telco embraces many variations on a theme, then the genus of communication provider contains many more. So you would think that, when asking for experience in RA, people might be more specific about what kinds of communication provider they want the candidates to have experience of. They rarely do, which either means that people from different backgrounds are either wasting their time or successfully competing for jobs that you might offer to people who have never stepped inside a communication provider before. I mean, how many years do you need to spend working in communication providers before you know enough about US GAAP to do a decent job of working out what can be accounted for as revenues? And why is IT development experience inside a communication provider so much different to IT development experience gained outside of communication providers?

Of course, the real problem is not with the candidates for RA jobs. It is with the people who offer the RA jobs in the first place. If only they knew what RA is, knew what they wanted, and knew what a good candidate looked like. To do that takes sufficient experience to write a good RA role profile and identify the appropriate candidate. The problem is many of them do not have that experience.

Chapter 13

Tangents on the Theme of Revenue Assurance

What Communication Providers Can Learn from Banking

Güera Romo

While working with a commercial bank in South Africa and Namibia, I tried to draw a comparison between revenue assurance (RA) in a communication provider environment and how, I imagine, it could be done in a banking environment. Take note of the use of the word "tried" here. I have to warn the readers that engaging with a commercial bank to roll out new banking infrastructure in Africa during these financial times takes some doing. There was very little energy left to daydream about RA.

Limiting this observation to the process of product and system implementation, what is the difference between implementing Amdocs, Singl.eView, and Flexcube? Not much except that Flexcube is implemented in the banking industry and not in a communication provider.

System Stability

A headache for any RA team is tracking usage through unstable systems. Inevitably the finger pointing ensues between two departments without either taking account-ability for system integrity. When the platform integrity was not assured during the implementation of the system, it is almost impossible to rectify this during production without at least several scars. The bank's implementation of the core banking system was no different. It did, however, make me realize how often the RA department picks up the accountability to ensure that system development staff is following project management 101 principles.

System Architecture

I have yet to work for a communication provider that has the entire system architecture documented in enough detail to follow the basic flow of data. The bank experience again was no different. I did organization development work for banks before, but this was the first time I was involved in banking operations. As a novice it would have been nice to work through the operating model down to a functional architecture and then consider how the system architecture would support or enable the business objectives. Only a small number of people understood what I was trying to do, and I was left again realizing the huge gap between technology and human capability. The gap just widens.

Solution Design

Many organizations, regardless of the industry, are forced to do solution design in parallel. This is achieved by continuing work in isolation with the intention of integrating later. The designs are based on assumptions and high-level directives. These usually come in the form of an MS Powerpoint slide presented to an executive committee. We build the solution from multiple points working in multiple directions and become confused when they do not tie up. This again is not specific to communication providers. Banks do it too.

Developing Specialized Functionality In-House

We are familiar with a few communication providers that are forced to (or elect to) design and build their own systems, be it an RA automation/verification tool or, strangely, a rating and billing engine. I am not an expert on banking payment systems, but I felt the business architect's pain when he wrestled with the concern of underlying functionality. Although he did not specifically mention complete and accurate billing, this is what he meant. The concern is independent of industry. The sheer size and complexity of interbank, intercountry payment settlement would make me think twice before embarking on the in-house development of a payment settlement solution. The closest I could come to offer help was to suggest we get a system auditor in, and that took some explaining.

Project Management and Communication

Coming back to project management, it is alarming to follow the demise of large programs (and those chunky budgets gone) due to ineffective project and program management. The governance and internal audit guys make plentiful visits to check up on the health of the program but somehow the message misses its audience or perhaps the audience has other realities simply not understood by reality-protected desk job executives—executives meaning glorified middle management without subordinates.

Product Development

Understanding your market and the appetite for certain products is vital to the costing and offering model. Similar to what we know as margin analysis, we would also look at the profitability of certain segments or go further to inform subscribers of more suitable packages to optimize their "bang for the buck." Products working for one country may not necessarily work in another country and this is not a function of gross domestic product or disposable income. There are nuances to what you sell, which can make or break the bottom line.

On the Upside

In banks, the volume of customers and complexity of parameters per product does not come close to that found in the communication providers I have worked with. Yes, we do have the difference in the size of countries I am referring to here but still, once operational, the speed and size of change do not require full-time RA people onboard. The company involved has an awesome fraud department with near real-time monitoring of transactions. This knowledge and infrastructure can be replicated into Africa. You can also train a businessperson in the basics of banking without having to turn them into technical superbeings. It is one of the few companies where procedure manuals actually mean something.

As to how they ensure that service charges and interest are calculated correctly, the answer was "I am not sure." It appears that the customer relationship management is adequate to know your assigned client accounts well enough to know if there is a problem on the account or not.

Personal Reflection

A famous quote in impoverished African countries is "complaining with a white bread under the arm." This refers to valuing or appreciating what you have. Here is an observation from someone starved of RA (such as myself). We often complain about working with communication provider executives who battle to grasp the basics of RA. We spend days dissecting their uninformed decisions and their inability to zoom in and out. When you are unfortunate in the nature of assignment the gods gave you, and you exchange the world of communication provider RA for something else, you come to miss those dumb executive decisions. You try for a while to marvel in the experiences of others, but somehow this is like trying to share a romantic meal for two, while your parents sit across the dinner table. Value the opportunities you have and help to build this RA profession into something that other industries will want to copy.

If Health Means Wealth, Then Plugs Mean Drugs

Eric Priezkalns

One difficulty with revenue assurance (RA) is the preponderance of metaphors that people use when talking about it. These metaphors may help with explaining the purpose of RA, especially in the absence of good public data to prove the point. However, all metaphors have their limits, and they can also lead to misunderstanding. Take the metaphor of a "leak" for example. Water goes in one end of a pipe, and comes out the other end. Not as much comes out as went in, because some is lost through leaks. Okay, but does this explain leaks caused by errors in rating? If water pipes could have rating errors, you might end up with more water coming out than went in…

Despite this, I am as guilty as anyone of using metaphors. In case you do not believe me, let me introduce you to one of my favorites from former US President J.F. Kennedy, who said that the time to fix the roof is when the sun is shining. If that is true, then an economic downturn should be worst for any communication provider that has not already invested in RA and less of a trouble to those that have. But I expect most vendors will be thinking the other way around "…look the roof is leaking and it's pouring with rain, so you had better pay someone to get up there and fix it now!" I think the problem with JFK's metaphor is pretty apparent. Any time is a good time to fix the roof. When the rain comes pouring in through your roof, fixing it will be harder, but that does not mean you can afford to sit and wait until the sun is shining again.

That is one way to interpret the metaphor about fixing the roof. On that reading, vendors would be roof menders, and if a communication provider has not mended their roof already, they had better get a vendor and do it now. But there is another way to apply this metaphor to RA. Now, I do not know about you, but if I paid someone to fix a roof, I would not expect that the roof still needed fixing on a daily basis forever more. If it did, I would probably argue that the roof had not been fixed properly to begin with. If the roof was fixed, there would be no leaks. I could sit snug and warm and dry, and not worry about it. I would not be walking around my house, looking up at the ceiling, and saying "where is the next leak going to come from?" I would not look up, find a leak, put a pan underneath to collect the drops, and start commenting to my family: "look at all this water we have collected—what a victory for our anti-leakage program!" In short, if you fix the roof properly, the job is at an end and you can stop worrying. There is no ongoing benefit, because there are no more leaks. You would not employ a team of people to keep looking up at the roof, ready to raise the alarm and run with a pan to recover the rainwater.

If we stick to this second way of interpreting the phrase, what would fixing the roof mean? It would mean building (or rebuilding) the roof so nothing went wrong. It does not mean waiting to see if something goes wrong, but spending a lot of time and money trying to deal with the consequences. A leak is, purely and simply, a mistake. Fixing the roof thus means having a business that does not make mistakes. A business that does not make mistakes has its decision-making processes lined up correctly, just like a roof with no leaks has all its tiles lined up correctly. Each decision fits the needs of the business. Whether a big decision (how should we price this new product?) or a small decision (where does the decimal point go in this new rate to be implemented?), the right decision is made. People take care to avoid mistakes, just as the roofer is careful to ensure the roof will not leak. Because individual people may not be reliable, you ensure the way they work is designed to avoid mistakes. That is what fixing the roof means to me—but it has nothing to do with what most RA vendors sell.

This week I read a news story about how obese people wasted a lot of money on products that misleadingly promise to help them lose weight, but which do no such thing. That got me thinking. There is a simple reason why the pharmaceutical industry is always looking for drugs to solve every problem that a person can have. Drugs can obviously be monetized. Whether the patient pays, or his or her insurance pays, or a taxpayer pays, somebody pays for every single pill. In contrast, prevention is not easily monetized. It is harder to make money from telling people to take some exercise, eat well, and take care of their body. It is easier to make money by waiting until people get sick and then stuffing them full of drugs. On the flip side, insurance firms and health-care services recognize that they will spend a lot less money if people try to be healthy, and spend a lot more money if they do nothing about it. It is worth remembering that big tobacco in the United States was eventually forced to admit the consequences of smoking because individual states had pursued them with huge bills for health-care costs.

Here comes another one of my favorite metaphors, and this one is especially for the RA industry: health is wealth, and plugs are drugs. A healthy business is a wealthy business, because the fundamentals of operations are sound. Mistakes are not made, so money is not lost. The job of RA in these companies is to promote health by promoting prevention as superior to cure. On the other hand, plugs are drugs. Leaks are a symptom, so plugging them gives instant relief. A business that spends all its effort plugging leaks is a business addicted to the drug of fixing a problem, but may do so at the expense of not addressing the underlying causes. The business is sick, but deals with the symptoms of illness instead of becoming well. The drug is addictive because there is an obvious and measurable benefit. However, life with those benefits is wrongly compared to life without those benefits. That is like persuading a sick man to keep popping pills because he will suffer more without them. Instead, the benefits of popping pills should be compared to the benefits of being healthy. The sick man should also look at the cost of his drugs and change his lifestyle so he does not need them in future. Prevention is better than cure. The question for everyone working in RA, whether for a communication provider or a software vendor or a consultant, is whether they prefer to sell health or sell a palliative for a chronic illness. Are we trying to make the difficult but virtuous sell of promoting the equivalent of healthy living for communication providers? Or are we happy just to push the drugs and make money by exploiting the sick?

A Memory for Revenue Assurance

Eric Priezkalns

I love the *LinkedIn* group that Morisso Taieb set up for revenue assurance (RA). For those unfamiliar with it *LinkedIn* is a social networking web site for people in business. Morisso's RA group in *LinkedIn* is the most successful public forum for debate about RA. However, there is one thing that frustrates me about it. Many of the questions being asked on the forum have already been asked—and answered— many times before. There are all sorts of public forums on the internet, and many of the questions seen in Morisso's *LinkedIn* group cover well-worn topics like "where should RA be in the organization chart?" and "what is the scope of RA?" which, although good questions, are going to have answers that do not change very often. It seems to me a shame that new answers keep being written from scratch, when good answers already exist somewhere on the Internet. Some of these questions have even been answered elsewhere on *LinkedIn*. Having a forum with questions and answers is one thing, but having some memory for which questions have been asked, and what answers were given would be a leap forward. Think of it as the *Google* principle: you can make the world a much better place not just by creating content—much of which will just replicate existing content—but by finding a way to search through the content that already exists.

Where am I going with this? I know that if RA does not develop a collective memory, it is doomed to ask, and answer, the same questions over and over. For example, if the *LinkedIn* group remains as it is, there is the danger that, a year down the line, new people will join, asking the same questions as previously answered. This will be boring and frustrating for people who have been active on the forum that whole time. This is one of the challenges for open group activities using the Internet—how to integrate new members while sustaining old ones. Some just close themselves to new members, or become so wrapped in their own world that nobody new feels welcome or wants to join. Others remain accessible to new members, but at the price of never progressing, which leads older members to lose interest.

So what is the solution? Perhaps a company like *LinkedIn* will keep on releasing more functionality, and that will help to solve the problem. A wiki for RA would help, but would enough people contribute to it? I could suggest that *talkRA* becomes a broader resource, but going beyond the existing search and archive functionality will be a lot more work. I am conscious of the extra effort it would take to aggregate answers from a wider collection of sources. Whoever does tackle this problem, the solution will be delivered in a closed group and hence there will be limited visibility of what they are doing, or it will be a public group that faces the never-ending problem of spam. Let us not forget that the *Wikipedia* page on RA was subjected to repeated spam and for a period was taken down because of it. Another problem is covering the cost of the effort involved in maintaining and managing any collective resource. *Wikipedia* depends on donations and goodwill from volunteers. A similar resource for a commercial study like RA will inevitably tempt people to manipulate it for gain, and even fair and reasonable advertising for any given vendor will likely discourage involvement from anybody representing a rival company.

So there you have it. I have posed a problem—how to generate a collective memory for RA—for which I have no straightforward answers. Does anyone reading this have a suggestion? I promise, whatever answers I receive, I will not forget them.

Can the ROC Be Lean?

Eric Priezkalns

Does Toyota have the equivalent of a ROC? It is a difficult thing to imagine, but a pertinent question to ask. Irrespective of their difficulties in recent times, Toyota is the most cited example of a lean business. In communications, the ROC is a concept that has become associated with, and sometimes dominates, the field of revenue assurance (RA). The ROC is often described as a component of a lean communication provider. But is the idea of the ROC consistent with being lean?

The term ROC was devised by Subex to help them market their products, though it could equally well be used to describe the suites offered by some of their rivals. The concept is analogous to the network operations centre (NOC). Put simply, the ROC aims to provide RA at an operational level by offering a centralized infrastructure to monitor the integrity of revenue streams, identifying and evaluating incidents so they can be resolved promptly. Another way of describing it would be the "mission control" for RA, where all the relevant data is not just collected but processed and presented so action can be taken to stop leaks. This nexus for automated extraction and analysis of data is designed to highlight where the issues are. It supports human analysts by presenting relevant data in a fashion that enables analysts to make effective and timely decisions about what needs further investigation and what corrections should be taken.

That explains the ROC. Now, what does it mean to be "lean?" The term was first used by James Womack and Daniel Jones in their book *Lean Thinking*

(Womack and Jones 2003). Their work was based on the Toyota model. The motivation for being lean came from the shortages of raw materials faced in Japanese car manufacturing after the Second World War. If they could eliminate waste, they would greatly reduce their costs. To do this, five principles were identified: specify what the customer really wants; challenge all wasted steps in producing a product that gives that to the customer; make production flow with every step adding value; pull from a subsequent step to a former one so flow is continuous; and focus management on attaining perfection by finding ways to reduce the number of steps, time, and information needed to deliver. The idea of lean is now very popular, and has been used in many other kinds of business.

In electronic communications, the organization that has most passionately promoted the concept of lean operations is the TM Forum. Keith Willetts, Chairman and Founder of the TMF, has even coauthored a book (Adams and Willetts 1996) and given tutorials on the application of lean thinking to communication providers. Though Willett advocates the adoption of fraud prevention and RA as tactics to improve efficiency, I believe there is an inconsistency between the concept of lean and the strategic emphasis that Subex places on the ROC. The ROC delivers retrospective quality controls through mass inspection. It does not aim to build in quality. Can the goal of the ROC be reconciled to the aim of being lean?

Although lean thinking was originally concerned with reducing waste in manufacturing, it is not hard to see how the thinking can be applied to business optimization within communication providers. Not getting all the revenues for services that have been provided can be seen as a kind of waste, in the same way that a leaking pipe represents a waste of water. Not only are revenues wasted, but there is also the wastage of effort expended in processing data that ultimately is faulty. Worse still, bad data may lead to bad decisions. Following the lean principles, it also follows that data is a waste if it serves no purpose. However, this is where we may start to question the applicability of lean principles to electronic communications. Lean principles can most obviously be applied to any operation where there is a physical or tangible sense of wastage, but the assumptions about optimal efficiency may not so readily apply when the product takes the form of a stream of data. A manufacturer may save a lot of money by reducing the wastage of raw materials. Saving on raw materials may significantly reduce unit costs and hence increase profits. In addition, wasted raw materials are not a useful source of information—other than as a measure of how much material has been wasted. Contrast this with data. Unlike physical material, the costs of data are not likely to be so readily correlated, or directly variable, with output. It has some cost, but it is not obvious whether targeting these costs would ever make a significant difference to profitability. Furthermore, it may not be obvious what data is needed to manage the business. There may be good reasons to build processes that generate or retain data that is unlikely to confer any clear benefits. Going one step further, the principles of being lean are often associated with gathering and using data to identify where processes can be made leaner. This idea underpins continuous improvement.

From this perspective, being lean is not consistent with being skimpy with data. This suggests that a data-driven ROC can be a valid element of lean operations.

One key way to tackle waste is to lower and ultimately eliminate the need to reject finished units because they fail quality control. A finished product that fails quality control represents a waste of raw materials and the expense that went into manufacturing the unit. The converse of a batch-driven quality check mentality is the idea of "right first time." It is an easy mantra to reproduce, but few in communication providers genuinely aspire to it. The business case justification of the ROC is to execute batch quality control. Instead of inspecting the quality of manufactured units, the ROC is used to inspect, via the intermediary of automated checks and reconciliations, the quality of data. In this regard, the ROC is actually a reversion to the batch quality model that lean businesses like Toyota had sought to replace. Their goal was to build in quality so that no units would fail. Building in quality means that batch checks find fewer issues and less resource is spent on reworking flawed outputs. To build in quality, you must break down a process into its elements, and try to make every element as simple and fool proof as possible, so there are no rejects later on. The microanalysis of processes needed to get things right first time runs counter to the mission control macro-overview obtained by the ROC.

To be fair to the ROC, the overused term "proactive" is often meant to signify the movement away from batch checking toward continuous or real-time checking. However, the way in which this is achieved often imposes a limit on how far it can go. If a check takes data from one system and compares it to data in another system, or somehow reprocesses the same data, the check still takes place after the fact. The delay before a batch check is executed may get shorter and shorter. The batches may be smaller and more frequent, to the point where each datum is checked individually. But, fundamentally, the objective still involves identifying and rejecting failed data, not preventing the possibility of failure before it even occurs.

The solution in reconciling the ROC to lean processes depends on what you see as the ultimate goal of the ROC. If the business case for the ROC is to identify flaws, and to fix those flaws, its mission is to execute retrospective quality control. This can be done in batches, or not; all that matters is the control is always after the error takes place. The more flaws identified by the ROC, the better the return on the investment made in the ROC. The logical conclusion is that a ROC, when viewed in isolation, generates greater returns if the processes are flawed, and not if they are fit to deliver data that is right first time. Conversely, if processes do consistently deliver data that is right first time, then the business justification of the ROC, from a RA perspective, cannot be to improve the bottom line in any methodical way. The justification must be demonstration of quality for the sake of demonstrating quality. Demonstration of quality may give peace of mind but will not lead to clearly measurable financial rewards.

In a business with a genuine "right first time" philosophy, and a successful approach, it would be hard to justify the expenditure on the ROC. While it might be useful to have an independent function confirm the quality of all processes, it is only useful if you have to assume that only an independent function could confirm the quality of processes. Otherwise, there would be no real advantage to having centralized checks of quality across all processes, when you could distribute the checks and building in of quality. Collating the quality control data in one place only adds further value if it is safe to assume that mistakes will be made, and that nobody will correct these mistakes without the intervention of a supervisory body. If checks were built into all processes and if people responded to them appropriately or if the processes were designed to be error proof, then the further supervision of the ROC would add nothing.

Where the ROC could add value to a lean operator would be in driving the decrease of error rates. However, the people employed in the ROC would ultimately be aiming to put themselves out of work, by driving the redesign of processes to avoid errors in future. The benefits of retrospective centralized quality control over streams of revenue and cost data could also be attained by implementing ever more elegant and detailed checks within processes, or better still by redesigning processes to eliminate the risk of error. Understood this way, the use of a ROC can be consistent with lean operations, but only when viewed as a medium-term tactical approach to identifying flaws, educating the business about quality and decreasing waste. The ROC can only be a strategic end goal if the business believes batch quality control and irregular rework of faulty data is more cost effective than making the investment to design and implement processes that will be right first time. Whether it is cheaper to find faults and fix them, or to eliminate faults before they happen, is not a matter of dogma or philosophy, but should be decided on the basis of empirical observation. Sadly, we lack the public data to know decisively either way, and we cannot form a general rule for communication providers that would be backed by hard evidence. But we can say that there are problems with trying to integrate the idea of a permanent and strategic ROC within the concept of genuinely lean telecoms operator in the way that a business like Toyota is lean. So the strategic question for communication providers can be put another way: is it leaner to design processes to prevent errors, or to allow errors to take place and fix them when they do?

Chapter 14

The Future(s) of Revenue Assurance

Future of Revenue Assurance?

Mike Willett

I was trying to think about how revenue assurance (RA) has moved, changed, and reinvented itself in the 13 years I have now been involved, in some form or another, with RA. In my view, there have been some movements, and if I look at those, then perhaps I can have some liberty and a crystal ball gaze at the future.

The first trend is the emergence of increasingly powerful software tools to specifically identify, address, and help manage RA. There should be little doubt that this has helped RA investigate new areas of revenue streams for potential loss while also automating much RA activity. This can be very beneficial for those new to RA who can draw on this expertise to learn from the experience of the vendor and deploy an RA capability quickly and usually with some financial benefit. The contrast of this is that the maturity of the tool set exceeds the maturity of the RA people, who may miss valuable steps in understanding the underpinnings of RA work and how their business works. The risk is that this can produce an overreliance on RA to be "solved" by the software tool. The greater risk, though, is that RA is "defined" by the software tool, its capabilities and roadmap, and not by the operator. Missing those early baby steps in RA can lead to a strategy that is aligned solely around the technology deployed and is hence, however well intentioned by the vendor, unbalanced.

This is similar to the issue faced by fraud management. In this case, I have seen many operators who struggle to adapt to new fraud types, until the vendor issues a software update, and who argue passionately that because their system does not detect that type of fraud, it is not really fraud. When the software update though is deployed, it can be thought of as a fraud again.

This leads to the second trend, which is prevalent in a growing number of organizations that provide RA services. If RA got into the limelight first during the dotcom and tech busts of the late 1990s and early 2000s then we should not be surprised that it has been reinvigorated through the latest global financial crisis. The mantra around billing or charging accurately for every event is particularly powerful when revenue growth is limited. But perhaps these service organizations are also seeking to fill a void in the operators where, due to the reliance on technology as discussed earlier, the ongoing return on investment from RA is not what was expected. But what might this void look like? First, as RA becomes more operationalized into the business, the RA team comprises a greater number of staff whose role is orientated around the RA tool. I have mentioned this earlier, but it leads to a distancing from the real data and business processes. More particularly, the ability to think about and challenge convention is reduced. This is not unique to RA, of course. Increased reliance on computing and the automation provided by vendors means loss of the real experts in a system or process. The imagination and insight of experts is replaced by staff who follow well-defined work instructions that ensure consistency, but perhaps not always quality. You could speculate that this is how RA was able to come into existence in the first place, as that level of complete and detailed knowledge held by system owners across the revenue chain was lost. This loss was not just from automation but also from the increased complexity that automation enabled. The risk from this move is that RA becomes more automated, and then the room for thinking disappears as reducing cost becomes an increasingly important corporate objective. And so, I expect that we will start to see and hear of an increasing number of examples of RA missing some significant leakage or undertaking poor quality work. In fact, this trend is already evident as I have had vendors indicate to me that when they have done a proof of concept at an operator, they have found leakage that the incumbent RA system missed. By the same measure, operators have spoken to me about an increasing false-positive rate, diminishing value of leakages identified, and their role being limited to raising the alert when finding a potential issue rather than raising the alert, managing the detail, and then helping in the resolution of the issues they found. This limitation of the role of RA is a by-product of RA only looking for the same leakage day after day, month after month, rather than expanding thinking to look into areas not yet automated.

The last trend I want to comment on is the move from reactive to proactive RA—however the words are defined. It is often said that the worst thing that an RA department can do is to set itself up as a profit centre, not least because of the impact it has on relationships with other functions like IT. This suggests to me that the creation of a function predicated around finding and recovering lost revenue can create a short-term star but one that burns twice as bright for half the life. For an operator, leakage should reduce over time. Complexity may increase but so should RA operational efficiencies; new products may be launched but so should more effective detection mechanisms; new business models may be introduced but RA people should know their own business and where the risks exist. And so, RA that built its business justification on leakage will find that it contributes diminishing returns and so investment becomes more difficult to justify. Looking forward then, I cannot see how any RA function can indefinitely continue to justify itself on the basis of recovery of leakage. The only question for each operator is how long it can justify RA this way. And so the move to the "Nirvana" of proactive RA—where RA does not have to find loss, it has to prevent it, and as importantly, the prevention of those losses are recognized as tangible and of business value. This is an issue RA must solve.

Against this backdrop, it should hardly be surprising then that RA people and vendors have sought to reinvent and legitimize themselves in many different ways. This includes aligning to more established functions, extending RA's remit beyond traditional switch-to-bill audits, moving into cost domains, moving from reactive to proactive, supporting transformation efforts, and seeking creditability through industry standardization. This article is not about my view on any of these but it is important to think on the motivation and rationale for any extension beyond traditional RA and understand in what direction, sometimes irreversible, this may take both the individual function and the overall discipline. The risk for RA is that it becomes too tool orientated and too operationalized such that it starts making errors (including errors of omission) while all the time returning less value. RA loses its attention to detail and understanding of the business and so becomes part of the problem not of the solution. Furthermore, the standardization of RA techniques will see the gradual removal of the strategic thinkers from RA teams and to vendors, consultancies, other functions, or other industries.

Having forecasted gloom, I believe RA still has the opportunity to add real and lasting value, but to do so it needs to address the following:

■ Developing software tools that expose data and its treatment to the RA function to ensure the end-to-end process is transparent and understood by RA
■ Having the different standard organizations define RA by the work that needs to be done and not by what can be done with tools
■ Defining "proactive" RA and a value proposition that extends beyond financial measures and ensuring this is communicated and understood at the most senior levels of the organization
■ Enhancing the alignment in RA between data integrity activities and process improvement to drive root cause resolution, and using tools and techniques already developed in these areas

- Extending RA and accepting its methodologies across other industries to allow cross-industry movement of RA people
- Learning on the part of the communication provider RA people as to how these challenges are met in other industries and incorporating that learning into best practices

Current and New Paradigms of Revenue Assurance

Eric Priezkalns

What is a paradigm? The scientist and historian Thomas Kuhn popularized the concept in *The Structure of Scientific Revolutions* (Kuhn 1962). He described it as a scientific framework embracing theory and philosophy. As a consequence, he observed that science does not progress smoothly and continuously, but does so in jumps, which he called paradigm shifts. When a shift occurs, one dominant paradigm is replaced with a new paradigm. A whole new way of thinking is adopted to explain observations and experimental results in the world around us. Kuhn's insight was that the current way of thinking is always contingent; it can be overturned by a new, better way of thinking. I wish to reapply Kuhn's conception of the paradigm to revenue assurance (RA), to highlight that there are two distinct paradigms in current use in modern RA. These two paradigms were independently formulated. It is a coincidence that both came to prominence and were described as "revenue assurance" at the same time. Despite their distinct roots, these two paradigms are to some extent reconcilable. However, a third paradigm is also emerging that may take over and become dominant.

To begin with, let me briefly describe the salient features of RA. By scientific standards, it is a relatively new discipline. It has not been developing long enough to be subject to a robust process of cross-examination based on the disparate views of its practitioners. Very few of the axioms of RA have been supported by objectively observed and publicly verifiable data. RA is characterized by

- A belief that all communication providers typically suffer errors
- A belief that the value released by resolving those errors is less than the cost of the solution
- An absence of data to corroborate these beliefs
- Disagreement about the exact scope of RA
- Disagreement about the best way to perform RA

In other words, there is relative consensus about why communication providers should do RA but competing theories about how to do it. None of the theories are backed by extensive corroboration or publicly available data.

The TM Forum's definition of RA identifies that the methods of RA fall into two distinct groups:

1. Data quality
2. Process improvement

As one of the proponents of that definition, two thoughts were at the front of my mind:

1. A definition is not based on personal feelings as to what words should mean, but rather the objective observation as to how people use the words in practice.
2. At the time the definition was proposed (and I believe this continues to be true) people used the phrase "revenue assurance" to refer to radically different approaches, with one camp concerned with auditing data quality and the other focusing on reviewing processes.

While some communication providers exclusively use one method or the other, both are legitimate ways to pursue the same RA goals. The majority use the data quality approach. A significant minority uses process improvement or a combination of data quality and process improvement.

That TM Forum definition was based on what communication providers were doing in 2004, and the definition has not changed since. However, a definition is based on observation of how words are used in practice. It is my contention that the practice of RA has changed since 2004 and continues to change. To explain how, we should begin with more discussion of the original paradigms built around data quality and process improvement.

Data Quality Paradigm

Figure 14.1 offers a very simple representation of the data quality paradigm. In the diagram, data is extracted from source OSS/BSS systems and then scrutinized by an RA analyst supported by RA systems. Issues identified by the RA analyst are passed to operational teams who then perform remedial actions on the OSS/BSS systems in response.

The properties of the data quality paradigm are as follows:

■ The RA team extracts data from source OSS/BSS systems. Collection may be more or less automated or manual in nature.
■ RA controls are executed by performing queries, reconciliations, and the like, on the data that was extracted. Analysts oversee this task. Again, the controls may be more or less automated.

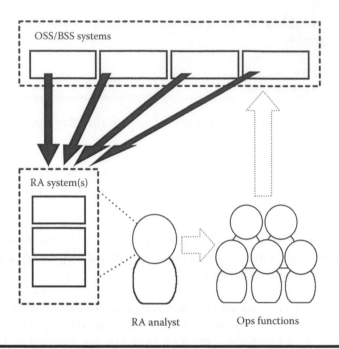

Figure 14.1 The data quality paradigm.

■ These controls are designed to find anomalies in the data, which following further investigation may prompt corrective action to recover historic leaks and/or prevent future recurrence.

■ When anomalies are found, it is up to the RA analyst to communicate relevant information and prompt action in other functions, predominantly those with a technology and operational responsibility. If the analyst does not communicate effectively with other functions, or the other functions are not adequately motivated to respond, the incidents will not be resolved.

■ Following communication with the RA analyst, other functions may further investigate and/or correct problems with the OSS/BSS systems that provided the source data.

■ This paradigm is synonymous with RA for many people.

■ It favors a technology- and data-oriented approach.

■ There is a "find and fix" mentality: look for problems, find them, and then fix them.

■ The approach focuses efforts on where it is easiest to obtain/analyze data and may ignore other issues where there is no available data.

■ There is a bias toward dealing with problems that involve large volumes of data rather than large one-off issues that can be diagnosed and remediated without needing extensive or specialized data analysis.

- The reliance on available data may lead to an incomplete coverage of controls, but can generate quick returns where data is available.
- Fixes prompted may be superficial and may not address root causes, leading to a recurring cycle of incidents that keep being found and addressed.
- Dedicated RA software used to automate the tasks may be difficult or costly to extend or modify to meet new requirements when the OSS/BSS systems change.
- Automated solutions may focus on what can be automated, which may not correspond to the real revenue leakage priorities.

Process Improvement Paradigm

Figure 14.2 presents the essentials of the process improvement paradigm. Unlike the data quality paradigm, RA practitioners do not query or mine data from OSS/BSS systems. Their skill is to engage with staff in the business, tasking them to perform their own checks and analysis of OSS/BSS systems, including checks of data. The aim of RA when following this paradigm is to understand processes and, by doing so, question staff in the communication provider about the possibility of process flaws or undetected issues.

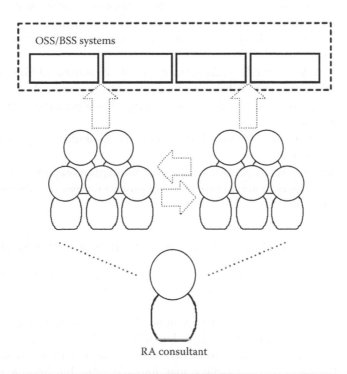

Figure 14.2   The process improvement paradigm.

The properties of the process improvement paradigm are as follows:

- The essence of RA work is to talk to people about what they do and how systems work.
- A checklist and process mapping approach is adopted to try to identify areas of risk, possibility for error, and potential gaps in the controls executed by the business.
- Understanding data flows is a key element in ensuring that RA reviews the processing of transaction data from end to end.
- RA consultants gather information from people in other business areas and task them to take action in response to their findings.
- The RA team will need to employ staff with a process and controls world-view and with the relevant skill sets and training required to put these into practice.
- Though this approach may be associated with traditional auditing, there is hardly any actual auditing because there is little or no inspection of data by the RA team. Instead, there is heavy reliance on the (possibly subjective) opinion of the RA consultant.
- The work of the RA consultant focuses on one area of the business for a period of time, and after covering that area and prompting change, they move on to follow the same approach with another area of the business.
- The technique can be slow and may fail to identify the quickest wins for the business.
- Because it does not rely on data, the methods can address all kinds of potential weaknesses, even if there are no actual leaks. For example, the methods can be applied to a forward-looking scrutiny of new products or transformation projects.
- The quality of the RA review is dependent on the skill and experience of the RA practitioner.
- This paradigm is less popular with most communication providers. This may reflect prejudices or actual experience of process reviews that did not lead to measurable benefits.

Mixing Data Quality and Process Improvement Paradigms

The strengths and weaknesses of the data quality and process improvement paradigms are complementary. As a consequence, some RA functions use both paradigms, and may effectively link the two together into a hybrid approach. For instance, the data quality paradigm is most efficient following investment into automation to extract and check data, but before making that investment, the process improvement approach can identify where investment is likely to be most beneficial. The data quality paradigm may find anomalies but finding anomalies is not an end in itself; the skills of the process improvement consultant are better suited to performing root cause analysis and determining the optimal solutions for the business. The data quality paradigm can deliver quantitative metrics, but does so by looking backward to problems that have already occurred. In contrast, while reliable quantification and prioritization is more difficult with the process improvement paradigm, its techniques can also be employed to think ahead and anticipate future issues, meaning it can be used at the design stage of projects that involve change. In general, more mature RA functions supplement the data quality paradigm with process work where there is no data. Some functions deliberately recruit two "wings" to the RA team, with the wings having the different kinds of skill needed for each paradigm.

Deploying both paradigms can lead to benefits, but it is also possible to link them. The RA maturity model integrates both paradigms at the higher levels of maturity. Though the two will typically evolve at a different rate, most usually with data quality being established and reaching a high level of sophistication, while process improvement is still primitive, the maturity model anticipates this. In the maturity model, the data quality paradigm alone is sufficient to reach midway in maturity, but then the techniques of process improvement are needed to keep moving up the scale. In this respect, the model reflects its parentage and ultimate debt owed to Deming's Plan-Do-Check-Act cycles and hence has a family connection to Six Sigma's Define-Measure-Analyze-Improve-Control cycles. While the data quality paradigm is good for study/measure stages of these cycles, it is weak at the plan/define/improve stages. In contrast, the process improvement paradigm is weak at measurement but good for other stages. By combining the two, a complete iterative cycle is created, with each paradigm serving to generate inputs to the other to drive continuous improvement, which is most fully realized at the "optimizing" level of maturity.

Emergence of a New Paradigm

Is there a way to get the advantages of both paradigms, but with fewer of the disadvantages of each, and while saving costs? A new paradigm is emerging within some communication providers. This new paradigm reflects wider trends in IT, most particularly:

■ Empowerment of users
■ Enabling flexibility without requiring the writing of new code

Moore's law, the observation that computing power and costs have improved exponentially over an extended period of time, underpins these trends. The new paradigm supersedes the data quality and process improvement paradigm by using better, cheaper technology to close the gap between the human skill sets needed on either "wing" of a hybrid RA function. Instead of employing people with process and interpersonal skills, but limited IT ability, plus people with consummate IT skills but no background in eliciting knowledge from people, the new paradigm seeks to enable the same RA practitioner to move seamlessly between both types of work.

The new paradigm is made viable by reducing the technology overhead for the RA function (Figure 14.3). Instead of utilizing specialized tools, which in turn are supported and used by specialized staff, multipurpose corporate business intelligence (BI) supports the needs of RA. In this regard, RA is just one of many customers of BI. Doing this opens two doors:

■ The repository of data that can be queried is no longer limited to what was uniquely fed for RA purposes; many more streams of data can also be accessed and factored into the work of RA.
■ A much wider range of staff throughout the business can interchangeably execute similar controls to those associated with the RA function, because they also have access to the common BI architecture.

The new paradigm becomes viable if detailed data can be made available, as it would be in a dedicated RA system, but without prohibitive cost and without intolerable delays in processing. The cost synergies of providing BI to multiple customers in the business will partly offset the investment needed in BI to achieve the high

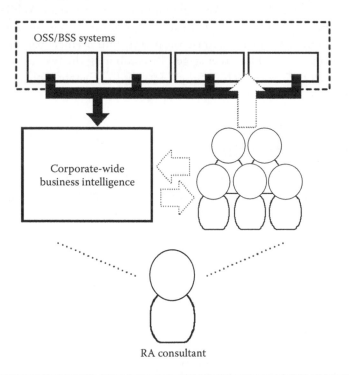

Figure 14.3   The new revenue assurance paradigm.

level of performance demanded for timely RA. Though such solutions, and hence the new paradigm continue to be the exception rather than the rule, more than one communication provider has now architected BI solutions that support RA performance comparable to, if not better than, that given by traditional dedicated solutions.

The main goals that drive the new paradigm are

- To allow staff skilled at interpersonal communication and process mapping the freedom to scrutinize data without also needing a high level of IT and database skills
- To thus increase productivity and flexibility of a limited number of dedicated RA staff

Other drivers for the new paradigm are to

- Reduce costs by sharing software/data/infrastructure with other business users instead of having dedicated solutions only for the RA team
- Reduce the development cost of RA automation by combining the generic best of breed technologies instead of writing similar codes that will only be used by the limited number of users in RA departments
- Allow skilled auditors to perform quick one-off queries so they can obtain quantitative data without their needing an investment in specialized, dedicated, and ongoing monitoring
- Avoid needing a mini-IT team employed by and working within the RA department
- Reduce dependence on support from specialist RA vendors
- Reduce time for the RA team to prototype new checks
- Simplify the handover of new controls to operational staff elsewhere in the business
- Increase the flexibility, scope, and role of RA by avoiding the need to set strict boundaries on the limits of the RA

Conclusion

It is too early to say if the new BI-led paradigm will supersede the data quality and process improvement paradigms that gave rise to the TM Forum's 2004 definition of RA. However, it has been realized as an alternative in some communication providers. This indicates that a Moore's law improvement of cost to power has opened new possibilities for automation of RA tasks. This can be successfully aligned to organizational structure and role definitions. A communication provider with a suitable BI infrastructure may wish to take an alternative approach to recruiting RA staff. Instead of expecting the RA practitioner to be stronger in either data or process-oriented skills, and balancing this by having a blend of people and working practices in the RA team, some of the IT-oriented demands on staff can be minimized, allowing process-oriented people to also work comfortably with detailed data. The lower premium on creating RA controls and queries, and the more extensive data available, also permits RA to take on a more fluid and extensive scope, while more easily transitioning responsibility for controls between the dedicated RA team and other functional areas.

Does Revenue Assurance Have Confidence Issues?

Mike Willett

Reflecting on industry debate, both on *talkRA* and more broadly, I found myself asking the above question.

My rationale was that much of the discussion around RA seems to center on what the next step of evolution for RA should be. It is almost as if finding and preventing leakage is not enough to sustain an RA function and that more needs to be added to the RA portfolio for it to continue to be seen as adding value. I wonder if the logic for this is an assumption that, with a well-established RA function, either leakage will fall away or, more likely, the costs of finding and fixing leakage will outweigh the benefits that it brings.

In many ways this makes sense. The RA maturity model advocates greater decentralization of the "RA" mindset and capability across an organization. Now in many organizations this is not practically possible to achieve but it is possible to move further along the continuum. And this would imply that the RA managers who do the right thing for their organization by spreading the thought processes of RA may find that their role reduces as resource moves out to operational areas because the perceived value of RA moves from tangible detection of loss to the less tangible prevention of loss.

So is it the case that if RA has done well, it means that the RA managers must look to extend their remit to hold their status in their organization? Is RA alone not "noble" enough to provide a sustainable career path? Now, I do not have the answers to this but as I have moved from working for an operator to being a consultant, this question is also relevant to me personally.

Surely, though, this challenge for RA is not that different to other risk management disciplines. Primarily, I am thinking of fraud and security, but this could also apply to other risks such as legal, regulatory, and so on. When RA is new in an organization, the opportunity for attention arises as the leakages are found and fixed, but at some stage a steady state will be met. Must RA extend itself to continue having influence and attention at the CxO level ... ? And, if not, will it find its position becoming less relevant? I do not have the answer to this question and I suspect it will actually evolve and change anyway over time, but it is something every RA manager driving his or her company along the maturity curve should consider.

Pedaling Profits
David Leshem

Recently I stumbled upon a story about Kenyans making cheap mobile phone chargers for their bicycles. What strikes me is the fact that one would expect it is in the best interest of a telco to *give* such a device to consumers. The logic is self-explanatory:

> no power = no calls
> no calls = no revenues
> no revenues = no telco and no revenue assurance (RA) as well!

No one could dispute that to have RA we must first have revenues. The cost of the charger is $4.50. Yet no one gives it away for free to the consumers. Amazing. By contrast, in Kenya the life insurance companies are providing free drugs to AIDS patients. These drugs will extend the life expectancy of their customer base and so improve their financial performance. The math seems pretty simple: as long as the person is alive he/she is paying the monthly premiums or at least not claiming on the person's life assurance policy, which improves the financial performance of the insurance company. It looks like the telco CxOs and their RA departments never thought about a parallel analogy in their domain. One comment though, as the CFO of one telco in Africa mentioned once, they provide free AIDS drugs to extend the life expectancy of their customers. It seems as though they did not think beyond that to also providing manual phone rechargers.

Of course, there are hand-operated phone rechargers. There are also solar rechargers. I even recall one top-notch finger-revolving recharger, but none of these come close to a $4.50 recharger to use with bicycles.

On a personal note, I thought that telcos already altered their engineering DNA by adopting the same techniques as mass consumer companies or copying how retailers conduct business. It seems like, at least for some, there is still a long way to go.

Content: Frontier for Revenue Assurance
Eric Priezkalns

They used to say that 3G stood for girls, goals, and gambling. That reduces the network to its purpose as seen by the customer. The network itself is uninteresting; what matters is the content it distributes. Communication providers have spent an extraordinary amount on technology so customers can watch video highlights of their favorite football team's last match while sitting on the bus. There is a gulf between what people *want* (to see a goal) and *how* they get it. In the modern era, how people get what they want can depend on so much. In the example, the customers' experience is built upon the use of better materials that make for a longer life of the tiny battery in their handset, and the management of the radio spectrum to enable the signal to be sent uninterrupted, and the use of remarkable mathematics to condense the content into the fewest possible signals while stopping people from listening in, and the cameraman's skills in following the action on the football pitch, and countless more besides. The operator is just one link in the chain of people and businesses that delivers the content to the person enjoying it. Customers do not need to care about how the content gets to them, just like they do not particularly care how a washing machine works so long as their shirts look clean and they do not particularly care how a train works so long as they arrive on time. What they really care about is that they never find themselves in a situation where they expected something, paid for something, but do not get what they expected and paid for. RA practitioners need to be clear on how far they want to get involved with content, and the real reasons why people pay for it.

In a sense, there is nothing new in the idea of communication providers distributing content. What I say to somebody else during a phone conversation is content of a sort. When people talk about content, they mean the kind of content where the same content is delivered to many recipients. In other words, they are talking about content that has more value because of the size of the audience. But the pricing of content is more sophisticated than that. Content can be valuable because many people want it, like the girls and goals of 3G. It can be valuable because some people want it a lot, and are prepared to pay a premium for it. A good example of that is business news and stock market data. Content can even be valuable when nobody wants to receive it, so long as somebody wants to send it. Advertising is the most obvious example of content that gets paid for by the source, not by the recipient. Communication providers face a question of how much they want to be involved in making money not just by providing a pipe that content flows down, but by exploiting the content that flows down the pipe. There is also a question of how much money they can make from content—both because of competition and because regulators will intervene to ensure there is competition. In the midst of this, RA departments need to look again at their mission. If the mission is to help the business make money—without limits—then RA will check not just that customers are billed for the content they receive, but that customers are willing to pay for it. If not, the business can suffer. The communication provider pays a cost for the content, and if the forecast of demand is wrong, the cost paid for content may be wrong too. This is not just a straightforward equation between the amount charged to the customer and the amount paid to the supplier. There is also the cost of working with a supplier (overheads like integrating a supplier to provide the content and for recurring processes like settlement) and the opportunity costs of providing the wrong content (customers will not search indefinitely for content; so, making it easy to find the wrong content has a cost if people find it too hard to find the right content).

All of these questions sound a world away from the concerns of RA. We expect marketing people, not RA people, to judge what kind of content might sell and for how much. Therein lies the heart of RA's dilemma. If the future of communication providers is to make a premium by distributing valuable content, and avoid the trap of selling only a dumb bitpipe, then RA is at risk of assuring only the dumb bitpipe part of operations. When businesses sell a customer a phone conversation by the second, and when customers pay for a conversation by the second, then it is clear why you need to get the charges right and receive the money for the service sold. When businesses sell girls, goals, and gambling, and customers pay for girls, goals, and gambling, business news, and advertising space, the relationship between the services sold and the charges rendered is no longer so straightforward. RA can try to step up and extend its role to really understanding what drives customer profits and satisfaction, or it can stick to assuring the bitpipe, almost ignorant of what people use it for and hence why customers pay, what they are paying for, and why they might stop paying.

The obstacle for RA here is one of imagination; not everyone will be able to make the leap and it is not clear if there are any executives who would even expect RA to do so. It would be easier to trust the management of nonlinear relationships between services and money to the kinds of people who feel more comfortable with them, especially the kinds of people who work in marketing. RA might then stick to what it knows well, which is the simple linear relationships between selling a unit and charging for a unit, be it a unit of time or a unit of bandwidth. The problem for RA is that only the dumb bitpipe business model is based on selling time and bandwidth. This means that while RA was once considered a "hot topic" it might end up looking like a relic found only in old-fashioned telco dinosaurs.

There will always be some value in assuring the bitpipe side of operations, but if profits are driven by content and how it is packaged, the value in assuring the bitpipe side of operations must diminish in significance relative to the business as a whole. This is what makes content, and the effect it will have on business models, the real frontier for RA. RA may return to what it was originally: a discipline for pioneers, who wrote their roadmap on a blank sheet of paper as they went along. Or RA may be penned in by the frontier, as it narrows around them, leaving RA doing a useful job in an area of reduced importance compared to the real drivers of profits and cash flow. Either way, the frontier is there and its importance is going to grow unless operators just give up on the content-led model and decide only to try and make their money from dumb bitpipes. The signs are that most communication providers fear becoming dumb bitpipe suppliers, even if they are not sure how to avoid it. In the meantime, the question for RA is whether it goes to the content frontier, or if it waits for the frontier to come to it.

Revenue Assurance Doublespeak

David Leshem

This article was inspired by a piece called *Life under the Chief Doublespeak Officer* (Lutz 1989). It was written by William Lutz, who went on to write a book about *Doublespeak* (Lutz 1990), and then several more. In my view, his writing about doublespeak is as relevant now as ever.

In this article, I would like to challenge the definition of RA, along similar lines to Lutz. I offer an idea why this function attracts only modest interest within a common telco.

No one would argue that the fundamental definition of RA is about assuring the communication provider makes money, this month. This necessitates that marketing, sales, and operations have a joint and collective role in managing a customer experience that delivers the brand promise. Yet, in how many communication providers do RA managers have the freedom to evaluate the above functions, without first going through a career change opportunity?

Marketing Defines the Brand Promise

Marketing identifies the most profitable customers and what their value is. Marketing documents and understands all aspects of the purchasing decision. It defines the attributes of product performance. It determines the brand promise and communicates it to the marketplace. In an ideal communication provider, marketing and operations work together to translate the customer experience into specific processes and actions through which the organization can deliver on the promise. If the brand cannot deliver on this promise, it is destined for failure.

- In how many telcos is RA an active contributor to these processes?
- In how many telcos does RA have any say in these topics?
- Would someone allow RA to review market research findings about brand strength?

Sales Effectiveness

Without a skilled and productive sales organization, few firms can survive, especially these days. The sales function takes place in a constantly evolving environment. Sales organizations must adapt to continuous changes in their products, customers, competitors, and markets. Intense competition places great value on understanding and responding to current trends within and across industries.

- What RA department would call up residential customers and ask their view on the communication provider's business?

- What RA department would pay a visit to a large corporate customer to learn about the issues that prompt them to take their business elsewhere?

Operations Creates the Right Infrastructure and Processes to Deliver the Brand Promise

The brand promise is only as good as the internal processes that deliver it. Every process should be designed, monitored, and evaluated on the basis of its ability to deliver against brand promise. This includes defining and removing internal obstacles and then strengthening the organizational "enhancers" (like communication systems and technology).

Successful organizational alignment means that marketing and operations have a joint and collective role in designing and managing a customer experience that delivers the brand promise.

- What RA department would ring up the call center to experience the promises that marketing states in its ads ... without risking being right-sized by the COO immediately afterward?
- What RA manager would propose metrics to evaluate operational effectiveness?

Conclusion

A typical RA function spends its time floating amidst a sea of troubles. These questions highlight how many troubles beset the communication provider, and how they are within the range of an ambitious RA function that has the right backing to really tackle them. The sea of troubles is wide and it is deep; I would rather see more RA functions taking arms and engaging in the fight to build a better business by tackling the manifold issues that undermine the fortunes of communication providers.

Evolution or Revolution

Eric Priezkalns

I must admit that I am a big fan of David Leshem's writing, and not just because he is part of the *talkRA* team. David is one of those few people who not only have the vision to see the world as it is, as opposed to how it might be, but also have the courage to say how they see things. His last article, entitled "Revenue Assurance Doublespeak," got me thinking. I was thinking about why I agree with David's insistence that communication providers should absorb revenue assurance (RA) into their metaphorical DNA, and why I doubt it will happen any time soon (if ever).

Knowing what should happen and being able to make it happen are two different things. We would all like to live in a world without crime, or one where nobody goes without food or shelter. However, that does not mean we can find or agree upon solutions to these problems. As an Israeli citizen, David will be very aware of how intractable some problems and conflicts can be, and how hard it is to reach an agreement about how to deal with them. On a more modest level, we would all agree that communication providers (CPs) would be better if they did not make mistakes. The problem is with agreeing on how to deliver on that goal, and its importance relative to other priorities for the business. I believe David, and others, draw a picture of how CPs would work in an ideal world. Today, I want to discuss the problems in plotting a course that takes us from this world to that utopia.

Suppose a CP has a certain way of doing things. Let us think of this "way of doing things" as the metaphorical equivalent of DNA—the code within our genes. Our real DNA defines us as living organisms, and the metaphorical DNA of a CP defines how people do their jobs. This code gets repeated from old employees to new employees. It is difficult to change because the code is repeated and instantiated by individuals, just as DNA is copied from one living cell to a new living cell. Changing the boss is not enough to change the attitudes of all the workers below. It may be that the workers change the boss more than the boss changes the workers. The bigger the company, the more the effort involved in making any changes. Individuals have a completely natural tendency to be conservative—people tend to stick with the things they know have worked in the past. If you did a job one way on one day, and you were not fired at the end of the day, chances are you will do the job the same way the day after. Even if individuals are asked to do things differently, they may have learned from experience that there may be more risk in making a change than in not making a change in the way things are done. The average employee may prefer the low-risk stance of avoiding change if there is a danger that change will lead to a screw-up and that screw-up may lead to them losing their job or denting their promotion prospects.

A gross simplification of David's analysis is that the corporate DNA of CPs is not open enough to getting functions to work together and confirm that their activities really do correspond to the best interests of the company as a whole. In David's vision, RA is more than just finding mistakes with retail bills. RA is about using system data and process review to align the behavior of disparate activities in very different parts of the business, to ensure they satisfy the company's overall goals (most pointedly, the goal of making money). That means RA—or whatever you prefer to call this holistic management of business interests—will have a relevant influence on many people across the company. As David identified, it will influence the regular activities of marketing, sales, and operations.

Holistic RA is a poor name for the vision David expressed, as his vision goes well beyond the boundaries of what many people mean by the words "revenue assurance." However, I will use it for now for want of a better phrase. There is no better phrase because we are mapping something that only exists in our imaginations, not something that we can routinely point to and show in real life. Lack of imagination is another kind of obstacle to change. It is a more difficult obstacle to overcome because people are unaware of it. So what are the mechanisms for change that turn something imaginary into something that is real?

Even if you can change the DNA of one cell, to change the whole organism you need to change the DNA of every cell. RA staff often talk about the importance of CEO sponsorship and engagement, but even a CEO cannot walk around a business and change the attitudes and methods of every single individual employee. It just would not be practical even if the CEO was god-like enough to know the details of how everyone might do their jobs better. If you sat down and created a new CP from scratch, then maybe involving David at the outset would help the business to adopt a way of doing things that is different to its competitors. It would include holistic RA within the corporate equivalent of its genetic code. Of course, doing that would still be hard. As soon as you give a job to somebody recruited from another CP, he or she may carry across the bad habits from the DNA of the former employers. Consultants are a little better, as they may have broader experience but still only experience how things are in the real companies they have visited. There is no guarantee that an external consultant is better at envisioning how to improve performance than an employee who manages the RA function.

To change a business we may need to alter the job specifications and work attitudes of a lot of people. They need to be molded and influenced to look at their jobs differently. Their sense of priorities may need to be realigned to the needs of the business as identified by someone like David. This in turn means the business must learn to reward the behaviors it really values. Because it is always easier to learn from experience, and copy something that actually exists, than to create something new, there are plenty of obstacles to realizing this form of holistic RA even if you started in a completely new company. Trying to change an existing company is magnitudes of difficulty harder.

Evolution is one mechanism for change. People often misuse the world evolution. Let me be precise about what evolution is. Evolution is not a change to an individual. Evolution is a change to a group. The DNA of an individual remains the same throughout its life. What happens to the group is that the proportion of individuals with different properties—different DNA—may change over time. Of course, a person's behavior is unlike his or her DNA in that it can change during the lifetime of the person, but let us assume that changing a person's behavior is hard to achieve, and there is no mechanism that will guarantee you can change any individual. Put simply, no matter what incentives, training, rewards, or demerits you use, you cannot be sure that any individual will behave in the way you want the person to. If somebody thrives in an environment, and you try to make the person change, there will be an obvious reason to resist—the person has thrived previously, so why would the change be an advantage to him or her? Even if *you* know how somebody else could do this person's job better, there is no foolproof method that ensures you can persuade that somebody else to change. I think this reflects the truth of the environment as confronted by people like David and I. We may think we know how the CP could work a lot better, but we lack the ability to persuade or force people to change accordingly.

There is an alternative to evolution. It is revolution. With a revolution, we find a way to change a lot of people very quickly. Any student of history, however, can tell you that revolutions are often bloody affairs. Many people may suffer during a revolution, but the revolution can also fail to deliver the promised utopia and be disastrous. The failure of past revolutions is yet another reason for individuals to be conservative and resist change. Another reason to discourage revolution is that, in any revolution, there will be winners and losers, so the likely losers will do what they can to prevent a revolution from taking place. Looking back at my own career, I am starting to believe that the kind of holistic changes promoted by David, other commentators, and I are just too revolutionary for the communications industry to adopt on any significant scale. It is not enough that *we* can see a path to a utopian future. The problem is that this path is barred by too many skeptical individuals, who have nothing to lose by staying exactly as they are, and may have a lot to lose if things change.

Imagine, for a moment, the kind of holistic CP that David envisioned in his doublespeak article. For a start, you will need to cross-check and cooperate with other functions and align with them. You may suddenly find your job has a lot less freedom. You may try to do an honest job of coordinating for the company's best interests, but find that other functions are blocking the right moves for the wrong reasons. You will need to work with hard facts—objective evidence based on what the data says about the company's performance—and not with personal theories. That can be tricky if you lack the necessary skills to understand the data. That can be dangerous if the "data" is in fact corrupt or has been misinterpreted. That can be embarrassing if the data shows your previous theories were nonsense and your past decisions were poor. Now consider that, when dealing with RA and related fields, we are largely talking about the propensity to commit *errors* in some sense. It should not be surprising to find only a minority of individuals are truly interested in making the changes necessary to improve business performance.

I sympathize and share David's frustration. There is a different way to run CPs, but it is hard to stay patient when others fail to, or refuse to see the changes that need to be made. A lot of theories about CP errors that used to be dismissed are now taken seriously. It used to be an exception to find a CP where solving RA errors was somebody's job. Now most of them have an RA team. The world did not change, so what changed is that people have woken up to a real problem. Maybe they had more imagination and realized the problem was there. Just as likely, they saw somebody else was dealing with the problem in another company so decided to copy them. Whatever the motivation to undertake RA, the work done by a typical RA team is only a fraction of the big picture that needs to be addressed to get optimal performance in CPs. It is only one piece in the puzzle. Most worryingly, once an RA team is established, they too may become refuseniks to change, worried about losing their jobs if things were done differently. I think both David and I would like to be revolutionaries of a sort—walking into a business, changing it completely around, then walking away feeling confident we left the business in much better shape than before. However, we know enough to realize how unlikely it is to stage a successful revolution, and usually they are only ever prompted when the business has already fallen into perpetual crisis.

I distinguished evolution from revolution, but evolution can sometimes lead to revolutionary change, if the circumstances are right. Take the poor old dinosaurs as an example. They were very successful for a very long time. They ruled the world. If everything had stayed the same, they would have remained at the top of the evolutionary ladder. Instead, there was a change in the planet's climate, and the dinosaurs were unable to adapt and survive. Warm-blooded mammals did adapt and did survive. Without the climatic transformation, the warm-blooded mammals called human beings would never have emerged. The same kind of thing can happen in business too. Businesses like Apple and Google pose a threat to traditional CPs. What is more, they are active in changing the ecosystem that provides the habitat for CPs.

Let us suppose the vision of a fully embedded, holistic RA can be realized in practice, and that the inevitable obstacles can be overcome. What is the likeliest path for its emergence? Will it be evolution or revolution? Evolution is slow. Revolution is fast, but happens rarely and is unpredictable. I think, if holistic RA is to be achieved, it will be the result of a combination of the two, much like that which occurred when the mammals took over from the dinosaurs. Many of the changes will be slow and evolutionary, but a change in circumstances may create a catalyst for accentuating the benefits gained from those accumulated small differences. Mammals evolved slowly, but the relatively sudden change in climate turned their warm-blooded adaptations into a pronounced advantage over their cold-blooded competitors. My guess is that truly holistic RA, of the type David imagines, will only emerge under similar conditions. It needs a CP that is not successful or large by current measures, but which is not under threat of going bust or being taken over either. This combination is vital to give it the extra incentive to do things differently to its rivals, while also preserving the long-term stability to really alter its DNA by mandating the necessary change at the level of each and every individual member of staff. There has to be some goal of being more than a successful business in one country, but of growing into an international business empire. The business empire will need sufficient capital to stay independent and grow by acquisition in turns, but not so much that it can afford to drop its disciplined adherence to efficiency, as it will always need to persistently maximize any commercial advantage over rivals. Staff will need to open to new ways of working and be willing to embrace a belief they can be world-beaters by doing things differently. Ambition is very important, as just copying a more mature overseas business will help a less mature company to close a gap on its betters, but will not help it to overtake them. Such ambition may be reinforced by cultural or national pride.

What might prompt the kind of change in climate that threatens the big beasts of the communication industry? We are seeing one possible change already, as vendors with a strong emotional link to customers, like Google and Apple, start to circumvent the importance of CPs, relegating them to commoditized carriers of (IP) traffic. Other changes in the climate are an increasing focus on the importance of content as a premium driver of revenues; mechanisms to deliver and share social networks and user-generated content, especially across multiple portals; intelligent marketing based on data gathered about users; and business models that leverage risk and creativity in the form of proposition-enhancing apps stores. All of these pose challenges to the business models of traditional telcos.

If I were looking at where in the world you might produce a CP willing to adopt holistic RA, the likeliest candidates would be multiplay incumbents in countries where the markets have been late to liberalize but which are now liberalizing, so there is a greater willingness to leapfrog to new approaches and rapidly embrace change. To encourage long-term investment in the business, growth should also be assured because of increasing standards of living, relatively low levels of service penetration to date, and a rising population in the domestic market. However, growth potential within the country should be sufficiently limited to encourage overseas expansion, particularly into neighboring markets. If those neighboring markets are already liberalized, this will encourage efficient working practices even if the incumbent is unchallenged in its home market. Countries that have commodity wealth will have an advantage in terms of relatively independent supplies of capital. Nations that already intend to develop their economy through strategic diversification into high-technology sectors like electronic communications will give a more favorable environment for any CP wanting to pursue holistic RA. The CP will need a long-term politically stable environment to pursue its objectives. This may also be coupled with national pride at expansion of business empires beyond its borders. All of this rules out Europe or North America as being the home of the first CPs to adopt holistic RA. Instead, it suggests that Africa, India, or the Middle East might be the likeliest home for the first CP run according to the principles of holistic RA. There is an analogy to the conditions that prompted Japan to become a world-leader in car manufacturing, though it will probably take at least as long—30 years—for such a success to emerge and for the reasons for success to become obvious.

Perhaps I have given David some hope with my analysis of the prospects of RA change being driven out of the so-called developing countries. Whether a 30-year period is best described as an evolution or revolution I will let you decide for yourselves. It all depends on how patient you are and how much time you can spare. I used to be very impatient for these kinds of changes to take place in CPs. Now I will be glad if we ever see a CP make it that far. When it comes to holistic RA, I believe we will eventually find a small but warm-blooded mouse that will beat the telco dinosaurs.

Chapter 15

Anecdotes from the Edge of Revenue Assurance

How to Make Bills Confusing

Eric Priezkalns

Now, I like to think I know a thing or two about telephone bills. So obviously I get upset if I struggle to understand my own bill. Let me tell you a story about my own home telecoms bill.

To begin with, let me set the scene. When I moved into a new house, being fond of simplicity, I did the simple thing and opted for the fixed line and broadband service from the same provider. By doing so, I would get all my charges presented on one bill, and have it all paid automatically each month from my bank account, by the magic of direct debit.

After a year or so, I was getting tired of interminable technical difficulties with the quality of my broadband connection and the provider's unwillingness to do anything about it. I decided to switch to another Internet Service Provider (ISP), while retaining my original supplier for my home telephone. The ISP transfer took place on the expected date of September 1. As predicted, when I tried to log on with my old provider that morning, their service was unavailable. Instead, I was able to connect to my new ISP. However, when my September bill arrived from my old provider, on September 6, broadband had been charged for the whole month of September at a cost of £21.27 before tax. Obviously, I rang up and told them to credit this as the changeover had taken place on September 1, meaning I should not have to pay for the service from that date. There was much discussion with the customer services representative (CSR) ("but it says on my screen" "well, it doesn't matter what it says on your screen because what your screen says is wrong"). The CSR insisted that the transfer had only taken place that day, September 6. Common sense finally prevailed as the conversation completed its fourth orbit, with the CSR concluding it was not worth arguing about. He agreed to credit me the charge for the month of September. I even got a nice letter dated September 7 saying the provider had credited my account by the £21.27 I had been overcharged for a month's service.

You can imagine how pleased I was when I got my October bill—which for some reason included a credit of only £11.18. This credit covered the period from September 15 to the end of the September. So, of course, I phoned my old broadband provider and complained again, this time about being charged for the first half of the month when I had already been promised a credit for the whole month. This time the CSR said the remaining credit owed to me had been put to my account—but I would only see it on my November bill. Perplexed, I waited for the November bill.

You can probably see where this is going. When I got the November bill, guess what? No additional credit. So I phoned again, the CSR agreed to credit the remaining £10.09 before tax, and I got a letter saying that the amount had been credited in the post the next day.

So in December, as anybody might, I looked at my bill to see the credit. There was no credit. At this point I was angry, but instead of cursing at some poor person in a call centre, I took a deep breath and double checked all the paperwork to see what had happened to my bill. I did this because, with each passing month, it was going to be harder to explain what had gone wrong to whomever I spoke to at customer services, and I wanted all the facts at my fingertips. However, I was surprised to find that after some detailed analysis, I could work out what had happened, without needing to have another call to the contact centre—but I doubt the average customer would have done the same. Checking the December bill carefully, there was a mysterious figure of £1.65 carried forward from November—but how could that be when my bank automatically pays my bills in full, through the direct debit scheme? Well, what had happened is that my provider, in their wisdom, had not taken any payment in November for the amount that was due that month. The amount due in November was £1.65 after taking off the credit of £10.09 (and grossing up for tax). So, because the provider had carried forward the £1.65 instead of taking payment, I actually got a higher bill than normal in December (I hardly ever make landline calls) but everything was even in the end. Everything was even, but because nobody warned me that the payment for a bill would not be taken as normal, and because there was no explanation that they had not taken the DD payment, I nearly cost the communication provider (CP) even more money by once again calling their contact center.

At least I could work out what was going on by looking at my paper bill. The same could not be said by analyzing my online bill, as I will now explain. Looking online to see what the CP's web portal said about my bills, I realized that the situation was more confused than ever. The online presentation of my bill made no mention of any credit. It just listed the unamended total bill values matching the bills that had been sent to me. Flicking to the payment history, November showed "part-payment received" which I guess makes a kind of topsy-turvy sense as the CP had netted the credit (a part payment of sorts!) but did not take the remainder owed by direct debit. But looking at the amount that was due in November, I found that on this screen it was *more* than the actual bill total (?) even though the actual bill total included the carried forward amount. The actual bill had been £13.50, but the screen said that £15.49 had been due. Hmmmm—how very mysterious. The carried forward amount, £1.65, was correct, so this implied a credit of £13.84 from £15.49 instead of a credit of £10.09 from £13.50. I looked around for numbers that could explain this strange anomaly of £15.49 − £13.50 = £1.99. And there was one number. No payment had been taken for my October bill either, because I was in credit at the end of the month. The half-month's credit I had originally received was greater than the amounts charged for that month. My account was in credit by £1.99, although to add further confusion the payment history on screen stated that £1.99 was owed. Of course, if you add the £1.99 credit balance in October to the further £13.50 credit put on my account in November, you get £15.49. So my old CP's online history, in short, only correctly stated one number—the amount due. All the others were fudged to get the final balance right.

Confusing, huh? You did well if you understood what I wrote. Now think about poor old me trying to work it out from the inadequate data given stated on paper or even more meager information given on screen. It seems hard to believe that a large CP can make such hard work of something as simple as crediting a bill and presenting the information to a customer. Well, it seems hard to believe unless you work for CPs and see what a mess they often make of things.

No Carry, No Cash

Eric Priezkalns

Author Douglas Adams created a plethora of memorable minor characters in his Hitchhikers series. One of my favorites is Rob McKenna, Rain God, who appears in *So Long, and Thanks for All the Fish* (Adams 1999). McKenna hardly appears in the stories, but he is easy to remember because he is a truck driver who is constantly rained upon. Wherever he goes, the rain clouds follow and rain upon him. Sometimes I feel like that, but with a telecoms twist.

I think of Rob McKenna whenever bad luck befalls my phone bills. Though it is not true, I sometimes think of myself as a god of billing errors. It is not like I am looking for them. On the contrary, I wish they were not there. It just so happens they keep looking for me. Unlike McKenna's rain and truck driving,

I guess being plagued by billing errors can be an advantage in my line of work. It certainly makes life easier, because it gives me lots of anecdotes and experience of what communication providers (CPs) get wrong. Of course, I would prefer just to get bills that are right. Knowing that errors chase me means that I have to scrutinize my bills very carefully, which is annoying and time consuming. Now, I had thought that we had cataloged every kind of leakage in the TM Forum's *RA Guidebook*. But I can proudly announce after an investigation of my own landline bills that lasted seventeen months (!) that I identified a completely new genus of leakage. I have never seen this kind of leakage before, and never heard anyone else describe a leakage of this type. If I were a Victorian botanist, I would hope it becomes known as *Priezkalns' Leakage Point*. However, I am not so sure I want my name associated with somebody else's mistake, just because I happened to find it. Instead, let us call this the case of the *indefinite ongoing carry-forward leakage*.

Back in December 2006 I first recounted the story of how my ADSL provider had made a mess of billing, crediting, and taking direct debit payments from my bank account when I stopped using them as an ISP. The conclusion of that story was that the CP had strangely decided to collect £1.65 too little in my November 2006 payment, and that the £1.65 reappeared on the following month's bill as a brought forward item. You will have to go back and read that story to understand why there was a £1.65 anomaly in the first place, because the explanation is an epic in its own right. Now, I thought that carrying forward £1.65 to my December 2006 bill would be the end of the story. I was wrong. In December 2006 my direct debit payment was again £1.65 less than the amount on the bill, and the value was brought forward to my January 2007 bill. The same thing has happened in every month since, with £1.65 being carried forward every month, and never being paid. Before you ask, yes I did check my bank statements every month, and every month the payment automatically taken from bank account was £1.65 less than the figure stated as owed on my bill. In March 2008 the same thing happened again, with my bill total including £1.65 brought forward from the previous bill, but the payment taken from my bank account was once again £1.65 less than the bill total. However, in April 2008, something changed. There was no brought forward line in the bill summary. The £1.65 had disappeared completely. It was nowhere on the bill. When payment was taken from my bank, it was equal to the amount on my bill for the first time since this had all started in 2006. For 16 months, my old CP had repeatedly postponed collecting the £1.65 that was genuinely owed, shifting the amount from one bill to the next. In April 2008, 17 months after the payment was due, it seems they gave up and wrote the amount off. No doubt the data was chugging around some interminable processing loop all that time, and finally reached an age where the data automatically expired or when somebody decided to purge it. Perhaps their aged debt analysis lacks the sensitivity to identify such a small amount, or perhaps it does not occur to anyone to monitor aged debt for customers who pay automatically from their banks and where the payments are always successful. Hence, courtesy of my old CP, we all have a new kind of leakage to look out for: indefinite ongoing carry-forward.

It has taken me many man-hours just to review my own bill. Compare that to the time spent each year by that CP's internal and external auditors on trying to work out if there are leaks in the charging chain. They might attempt to audit a mediation platform in no more time than it took me to write this article. The sad truth is that consumers are better off relying on themselves than on auditors. If forced to make a choice between relying on Rob McKenna or the typical CP auditor, I would rather take my chances on a driving holiday with Rob McKenna, Rain God. At least we could share stories about our bad luck (or rather, our dismal destinies ...).

When Revenue Assurance Stops: A Bedtime Story

David Leshem

Many years ago, in the year of 2001, there was Genie, the mobile Internet portal by BT. Genie was an attempt by BT to unify various global brands into one, as well as to address the mobile Internet. In the United Kingdom it was BT Cellnet, in Germany VIAG, in the Netherlands Telfort, and in Ireland Digiphone. Nowadays after change of ownership it is known as O2.

When I met the chief data and marketing officer of O2 and the president of Genie, a long time ago, his vision was clear: data is data, and kilobytes are kilobytes. Just as customers pay the kWh cost for the electricity used to power their fridge or TV, mobile Internet customers, he believed, should pay for the data they consume. To make life simple, the cost of each KB should be the same, whatever the customer is using it for.

I was hooked. This felt so good, like the feeling I get before making a smash in tennis. Yet there was some degree of doubt. So I asked him softly, if you are going to charge for a banking balance query say £0.80, and it usually comprised of 80 bytes, that means 1 byte costs £0.01. Would that mean when I download a 100 kB web page, I should pay £1,000?

You would agree there was no point to continue the discussion. After all I was a vendor, and he was a potential customer and a popular keynote speaker. So even though it is a vivid case of a fundamental flaw in the business model, it was not my role to confront him with blunt questions. The end result of Genie is history.

The one person who made money from Genie, and similar failures along the way, is Matt Haig. He wrote *Brand Failures: The Truth about the 100 Biggest Branding Mistakes of All Time* (Haig 2003).

Let us take a step back and discuss the communication provider business as a business with the objective to make money, or at least not to loose money. Whose role it is to make sure that the marketers come up with the right tariffs? As a good and old friend of mine once said (when he was still a billing manager), "every time the marketing guys get drunk, I have a new tariff." I would like to think we can all agree that, when the tariff is wrong, and costs the business money, finding leakage related to the tariff is less important than understanding why the business chose a bad tariff to begin with. I am quite puzzled by the fact that I cannot recall a single communication provider where RA helps evaluate which tariffs to launch and which tariffs to scratch.

Battlegrounds and Graveyards

Lee Scargall

There is a well-known saying at Cable & Wireless that goes, "Jamaica is the battleground and also the graveyard for Chief Executives!" When CEO and President Rodney Davis exited the business, he was the fourth CEO to leave within the 7 years since the Jamaican market had been liberalized. He was preceded by Errald Miller, Gary Barrow, and Jacqueline Holding. Their fate came to mind after reading David Leshem's very interesting article entitled "Where revenue assurance stops, a bedtime story." The article reminded me of what happened when Davis was at the helm in Jamaica, when in a desperate attempt to claw back market share in both the mobile and fixed-line businesses he introduced a mobile tariff called "10/8" and also a pre-pay plain old telephone service (POTS) service marketed as "homefone."

Davis was eventually axed after a disastrous first quarter in 2007, posting significant losses owing to both of these products. In the case of the 10/8 mobile tariff, the profit margin was so low and the volume of off-net traffic to local competitor, Digicel, was so high, the tariff was always going to be a problem to sustain. In the case of homefone, the pre-pay POTS service was positioned in the market with free installation, which had disastrous consequences.

The homefone service was in fact hugely popular amongst the low-income earners in Jamaica, and over 120,000 new subscribers applied. After putting in new line plant to accommodate growing demand, the return on investment would not break even until year five. Most homefone customers could not even afford to make regular calls anyway, but at least they all had shiny new phones, and some even had multiple phones in each household for good measure. After all, everything was free, free, free—even free installation! Subsequently, opex and capex lines soared without any material increases in revenues and brought Davis' tenure to an end.

This brings me back to the question in David Leshem's article: "Whose role is it to make sure that the marketers come up with the right tariffs?" Well, I often ask myself this question over and over because I was head of revenue assurance (RA) at Cable & Wireless International at the time that Davis was introducing 10/8 and homefone. Perhaps I should have done more? Could I have done more? Who else in the business was noticing this? Why do we continue to provide these services when there is no payback? Why are these tariffs so low? The questions would continue long after the event.

As RA practitioners, I believe we are here to inform the business that these things might happen. The problem is that when you have a desperate CEO trying to make his numbers before his final showdown at the O.K. Corral battleground, then all rationality (including the advice of the RA team) goes out of the window. As it happens, Davis was not the only casualty in all this, and the axe later fell on Cable & Wireless' entire London headquarters.

Abstraction Reactions

Eric Priezkalns

In the electronic communications industry, time is money in a very literal sense. Even those with the most basic understanding of communication providers know that. In one of my former jobs, I fondly remember tormenting an audit junior who was tasked with the important job of identifying weaknesses in my revenue assurance (RA) strategy. She kept stressing the importance of not losing minutes, so I kept asking her where I would normally find these "minutes" she kept referring to. Poor girl, she lacked the training to even begin to formulate an answer. There was no talk of protocols to transmit files or definitions of the fields in call records. I dare to think what would have happened if I resorted to the old and detestable trick of throwing some impenetrable jargon and acronyms into the conversation; if she did not know what a CDR (call detail record) is, then talking about AMA (automatic message accounting) would have left her nonplussed. But I did not play that game. I just kept asking her questions that were technically precise, when I knew she did not have the technical fundamentals to respond in kind. She kept saying that minutes could be lost, as if minutes were sheep and I was the shepherd. She simply did not know the subject matter, which made it easy for me to give her a hard time.

Do I feel bad about being mean to this poor audit junior? Not really. Time is money, metaphorically as well as literally, and communication providers pay for the time of their auditors, and for the time of staff like me as well. Her ill-considered questions were wasting her time, but not wasting as much of my time as a few ill-considered audit findings would. When an audit junior comes after you asking about minutes being lost, but unable to formulate any idea about reasonable risk, or even if the topic bears any relation to the purpose of the audit, the potential for time-wasting audit findings looms large. If I had really wanted to tear the audit junior apart, I would have pointed out that her questions could be best described as value-add because there was no connection between them and the reliability of the financial accounts (if you lost data on transactions, it would be prudent not to recognize the revenue, hence leakage may impact the final revenue number, but does not impact the integrity of the final number). And there is not much value added by an auditor who asks questions without being able to understand the answers properly. Even so, I finally relented and took pity on the poor girl, explaining what CDRs are and where they come from, so she could write up her management letter points and bluff a little more persuasively at the next client she visited.

Part of the problem with telecommunications operations, like any complex subject matter, is that the abstractions people use to understand the topic are not going to be consistently reliable. The poor audit junior knew that losing minutes was bad, but had no idea that the minutes she was talking about are an abstraction based on the difference between two data fields—start time and end time—and that a record of a call is itself nothing but a stream of data more or less bulked up into files of whatever format on whatever disk space associated with whatever computer. Not knowing that, the audit junior was obviously going to struggle to go into detail as to how these minutes might be lost. Her abstract model knew that minutes started somewhere, and ended somewhere else, but she had no idea of what they were or how they might go missing. She also was going to find it hard to conceive of the many ways that minutes could go missing. She could understand losing or corrupting data, but would always have conceptual blind spots when it came to the difficulties of knowing whether you recorded the start and end time accurately to begin with, or the cumulative loss if you chop off some digits in the trade-off between processing burden and precision.

For me, abstraction is the single biggest problem for RA. Lots of people want to talk intelligently and confidently in terms of abstract representations of what can go wrong. The problem with most of those abstractions, even if they are insightful and given by someone very confident (unlike the poor audit junior) is that the abstractions will ultimately break down in the face of the crushing detail needed to execute a real RA control. Abstractions may obscure vital aspects of what is happening in the detail, or skew the perception of risk so that effort is misdirected. I have been in more than one communication provider that has put disproportionate effort in assuring usage compared to nonusage because people could imagine what might go wrong with usage but not with nonusage. The same bias also occurs between postpaid and prepaid; communication providers that earn most of their money from prepaid services sometimes pour a disproportionate amount of resources into assurance of postpaid billing simply because it is easier to work out how to assure postpaid billing. Similarly, when checking usage, people tend to put disproportionate effort into the problem of getting data from one place to another, and forget about problems like assuring whether the data was any good in the first place. Abstraction is vital for communicating and for having an overall understanding of the domain where risks are treated according to priority, but poor abstraction leads to poor decisions. Like so many things in life, the devil, with RA, is in the detail.

There are layers and layers of abstraction to electronic communications, of course, and I use the word "layer" deliberately there. Knowledge of detail is essential, but knowing the detail of the layers in the open system interconnection

(OSI) model, or the very different kinds of layers in the Committee of Sponsoring Organizations of the Treadway Commission (COSO) framework, or of the layers of hierarchical process decomposition in the enhanced telecom operations map (eTOM), is only going to get you so far. In the end, they are all useless unless you can also understand the connections between those abstractions and the systems and practices in place at whichever communication provider you are dealing with. I remember an encounter with another group of auditors, who were employed to check the accuracy of retail billing for a communication provider. They spent a day asking their questions to whatever poor souls they had decided to pick upon. At the end of the day, they came to the conclusion that they had serious concerns with the precision of the records used. That resulted in some serious concerns on my part—about the competence of the auditors. After a couple of pertinent questions about which systems they felt to be at fault, it transpired they had wasted the day reviewing the generation of records only used for interconnect billing—which was completely outside their scope and unconnected to their audit of retail billing accuracy. Although the auditors made a gross error in performing their work, on one level it is easy to understand. They got so tied up with the detail of the systems they were looking at, they forgot to ask themselves some basic questions about how those systems related to their audit goals.

Detailed knowledge is fundamental, but no individual is going to have enough knowledge to cover all aspects of RA. The trick is to be able to apply a useful abstract model, and to bend and change it to reflect the real and detailed facts. That requires professional skepticism, not only of the statements presented as fact, but also of the abstract models used to organize those facts. The two are in constant tension. The abstraction may lead to questions about the details, and the details may prompt questions about the suitability of the abstraction.

There is another aspect to abstraction that people sometimes skirt around or conveniently ignore. Abstraction goes deeper and higher than just the technical detail of how systems and processes work. If you want to get really skeptical at the detailed end of RA, you end up needing to understand things like the fetch-decode-execute cycle of computers or the speed of electrons fizzing down a wire. Trust me, there is a lot you can learn if you want to keep on pushing RA to an ever more detailed level; for example, I once read a paper on the effect of weather conditions on the speed of radio waves. I have seen people talk, at a supposedly detailed level, about how things like signal times from handset to switch are constants for communication providers, when a moment's thought indicates they are not. They are not constants, so the question is how variable they are.

At the other end of the scale, we must not lose sight of why we do RA. Somebody needs to see a benefit, be it customers or shareholders. Yet many abstractions of RA simply come up with some hideously and unjustifiable assertion about what all customers and/or shareholders want, or worse still, pretend that all stakeholder goals can be neatly and conveniently aligned. Only a simpleton (or somebody not stepping back and taking a moment to think about what they are doing) actually believes that the priorities of all stakeholders are identical and are equally satisfied by the same choices. In all choices there will be trade-offs, and an overly simplistic opinion about what stakeholders want will only lead RA people to persuade themselves of their success, while failing to recognize the compromises they have actually chosen.

Building up RA abstractions to link the most excruciating detail with the most pervasive understanding of stakeholders is a never-ending process. It requires perpetual addition and refinement. The irony is that RA is about spotting and stopping errors and wastage, yet without the right abstractions, you may not be able to communicate about them, or even conceive of them. Then, armed with the right abstractions, they must be applied to the mundane but fundamental detail of how things actually work.

Metering Not Billing

Eric Priezkalns

It seems amazing to me, but I hardly need to do research to find out how poor businesses are at issuing bills. I just need to open my own personal mail to find that out. Sometimes it is good that billing errors are mysteriously drawn to me. It was several years after moving into my newly constructed house that my duel-fuel gas and electricity supplier finally issued me with a bill. I always had every intention of paying, just no intention of wasting my time chasing my supplier to ask them why they had not billed me before.

In fact, until I received the bill, I had no idea who my gas and electricity service provider was. The one time I had a temporary problem with the electricity supply I spent two hours on the telephone trying to find out whom to contact, with no success at all. In the end, a man from the construction business responsible for building the estate came around knocking on people's doors, telling everyone why they had been pitched into darkness. His explanation was succinct—it was all the fault of another business. I thanked the engineer for taking the trouble to let me know my electricity was off (something I knew already) and that there was nothing he could do about it (something I did not know, but of little use to me). There did not seem much point trying to complain further, and stoically accepted that I should always keep a store of candles just in case, as well as knowing where the torch is. It says a lot about how these businesses work that a diligent and caring person walking door to door still ends up being the quickest and most effective form of communication when something goes wrong. And that it still leaves you completely impotent to do anything about it.

So why am I writing about this? Because I struggle to understand how my utility service provider failed to bill me for so long. Ever since I moved in, people with official badges had come around on a regular basis to read the gas and electricity meters. All the meter readings were stated on my first bill. So why did it take them so long just to reconcile the meter readings to the bills? It should be a basic control to check that a bill is issued for each meter reading taken. After all, paying people to read meters is a real and significant cost and there is plenty of attention paid to the problem of balancing the cost of taking meter readings with the issues involved in estimating the amount supplied. I will not mention the name of my supplier. However, if I were a shareholder I would be wondering how much extra cash would be generated for dividends if they simply managed the unbilled meter readings they receive each quarter.

Saving Pennies, Saving Face

Eric Priezkalns

On *talkRA*, David Leshem asked a difficult question about which countries appreciate revenue assurance (RA), and which have little interest in it. This started me thinking. One of the difficulties in comparing RA is that a communication provider that seems to have no issues may have an RA team that is not reporting them, and a communication provider that seems to have lots of issues may be benefiting from a hardworking RA team that diligently finds and reports all these issues. It comes down to culture. Pride can be a powerful motivator. On the other hand, the need to save face can stop people from admitting and dealing with the real issues they face.

I remember one relevant story from my early days as a financial auditor. I worked on an odd mix of clients, including a Japanese travel agency based in London. The travel agency serviced a variety of Japanese corporate clients also based in London. Like accountants in most small businesses, their accountants focused on the day-to-day essentials and let the auditors worry about the adjustments needed for the year-end financial statements. Come the end of the financial year, there are always some transactions where the bills have come through the post, but nobody has got around to paying for them. You need to be careful that some do not get missed. To check, you ask the accountants for all the recent bills they got in the post, pick a sample, and see if they were included in the totals. We asked the accountant at the travel agency for the paperwork as usual, and he told us which pre-prepared file he had put it in. After coming back from lunch, and quickly reviewing the invoices, it became obvious there was something wrong. There were only a few invoices in the file, yet a travel agency deals with lots of transactions every day.

On the accountant's return from his own lunch, we asked him to come into the windowless room where we were working, and proceeded to explain what we needed. He insisted that all the paperwork had been provided already. The accountant was Japanese, like everyone else at the business. We thought there had been a misunderstanding, perhaps to do with language, so we explained again. Again, he insisted. So we tried to explain why we expected to see a lot more invoices, and he shook his head and said we had them all. It was hard to work out if the problem was to do with his comprehension of our English, so we asked again. His reaction made it hard to tell if he really did understood, but he said there were no other invoices. We went on like this, over and over, for the rest of the day. We knew the invoices were too few in number, and it seemed unlikely the accountant had deliberately made an error that was so easy to spot. We asked again and again, and over and over he, very calmly, slowly, and quietly, said he had provided all the invoices. Come leaving time, we shrugged our shoulders and decided to give up, while not entirely sure how we were going to handle what seemed to be a serious obstacle to checking the accounts.

What were we doing wrong? In my ignorance, as a junior auditor, I failed to consider cultural differences. Keeping the accountant in the room, and insisting on the invoices, was my mistake. I should have realized that language was not really the problem. The problem was with the threat of losing face. From my limited experience, British accountants would provide you with paperwork when you asked. They give it when asked because they expect you to ask; they are less likely to have everything prepared in advance of the auditor's visit. However, this Japanese accountant took it as a point of pride that he had given the auditors all the paperwork they needed at the outset. While we thought nothing of the fact that some invoices might not have been provided until we asked for them, he felt at risk of losing face. Half of that day had been spent going over and over a pointless conversation. When we returned to work the next day, there was a new object waiting for us in our windowless room—a shoebox full of unfiled invoices, all with recent dates on them. The accountant had understood what we needed; he just could not bring himself to get the box while we were asking for it. He could leave the box for us, so long as he did it when nobody was looking.

The Japanese accountant was diligent, and did his job well. He did not want to lose face, even if it seemed to us like that there was no reason why he would. Diligence is vital to good RA, but pride can sometimes cause obstacles to communication. There are no easy generalizations, but attitudes do vary with cultures. When it comes to judging how good RA is across the world, we need to listen to what people say, and what they do not say, and also understand what motivates them to talk and what causes them to be silent. This will be especially important to RA, which is concerned with the inadvertent loss of revenues, and the action needed to save them. We need to understand a person's motivation to understand what they say and do. Only then can we judge their real attitude to RA.

Chapter 16

It Makes You Think ...

Hidden Switch

David Stuart

I am going to describe a very bizarre situation. A mobile operator I was working for rented a hosted switch. As part of the deal, a failover switch had also been provided. The operator had only set up CDR collection from the production switch, as it was believed that the failover would be dynamically allocating the IP address of the production machine in the event of a failure. No issues yet; however, due to a misconfiguration/misunderstanding at the hardware provider, the failover switch had not been configured as a failover, but as a second operational switch instead. As a result, 50 percent of the operator's events were being sent via this second switch, and hence the operator was receiving only 50 percent of the CDRs and therefore billing only 50 percent of the events.

There are a couple of ways by which this type of revenue loss could be identified. So the question is how to identify this loss?

* * *

The revenue assurance (RA) solution that flagged up the missing switch problem is the oldest and probably one of the best—I checked my bill! I kept a log of all events and calls I made from my mobile, and once a month I checked them off against an internal bill. Once I found a number of events missing from the bill, and I had to perform the hard part of RA—the root cause analysis. After trudging through suspense buckets and mediation dropped records, and being unable to find the CDRs, I conducted a couple of hundred test calls. From the results of the test calls, I could see that the missing calls were completely random. This led me to work with the people in the engineering department—who were completely convinced the issue must be IT related. We conducted test calls while monitoring the known switch—the first three events were completed successfully and were all monitored on the switch; the fourth event was also completed successfully, but at no point did it show up on the switch. The engineer of course declared that this was an impossibility, but further testing proved that only about 50 percent of events were appearing on the known switch. An urgent meeting was organized with the hardware vendor and the issue was uncovered.

By checking my bill, I had managed to uncover a very significant revenue loss. The only other methodology for finding this loss would have been to use test call generators (TCGs). A TCG operates in pretty much the same way as my handset and I—just a bit more automated and a lot more expensive!

Fraudulent Engineer

David Stuart

While working for a fixed line operator, the following scenario was discovered. A single switch that served half of the country was found to have nearly a thousand customers on it who were not set up in the billing system. The customers had been set up in such a way that whenever they made calls they did not create CDRs—so there was no downstream issue with suspense. After a lengthy investigation, it was found that the engineer who administered the switch had offered the residents in the local town an amazing deal of free calls for life, for a one-off fee—which he pocketed.

Although this is actually a case of fraud, it was discovered using RA methodologies. There are numerous ways by which this issue could be discovered, but how would your existing controls/solutions identify this loss?

* * *

The purpose of this question was to demonstrate that certain scenarios can be picked up by numerous RA controls, all very valid, but varying in degrees of complexity and expense.

I found this revenue loss in a good old-fashioned operator, where CDRs were created for every scenario—these are less and less common as IT managers attempt to reduce costs by insisting that only billable CDRs are generated at the switch. Switch A was where the fraud was occurring and Switch B represented the other half of the country. I quite simply isolated all of the terminating CDRs from switch B where the calls had originated from Switch A and reconciled them to the originating CDRs (billable) from Switch A. At first, the query was performed at the volume level (a five-minute check) but due to the disproportionate volumes (in both directions), a detailed reconciliation was then performed. This flagged up some numbers on Switch A that did not seem to create CDRs, so the customer accounts where then checked and found not to exist. From here, a full investigation was then conducted and the root cause eventually found.

The above reconciliation was not the best way to find the problem as it was dependent on the individuals making calls. However, all it needed to do was to find one problem and from there a good root cause analysis should lead to the full picture.

VAT Man

David Stuart

This question actually comes in two parts. The usual—how would you identify this loss?—and a second question: Does this fall under the remit of RA?

Consider the following scenario: In the United Kingdom, VAT on mobile phone calls is calculated by one of two ways: for calls originating within the European Union, VAT is charged at 17.5 percent (the standard UK rate), and for calls originating outside the European Union, VAT is zero rated. The operator I was working for had assumed that IT systems were calculating these figures accurately and were paying the VAT man accordingly. However, this was not the case and the operator was in fact paying 17.5 percent of all prepaid call revenues to the VAT man. Needless to say, the financial implication of this was massive.

I have slightly simplified the above scenario—as I do not want to delve into the legislative joys of VAT, but hopefully I have painted the underlying picture.

* * *

I do not believe that this particular issue can be found using any conventional RA reconciliation. This issue was found by simply talking to a colleague in the finance department.

At the time, I was responsible for RA compliance of all new product releases. Due to staffing reduction I was also asked to take on the responsibility of ensuring that all of finance's requirements were incorporated into new product releases. As a result, this meant sitting down with the VAT manager and defining his requirements for a new call feature. During the discussion, he explained the rules for VAT calculation and at this point I drifted off, trying to work out how what he told me was possible. I knew that there was a flag on the postpaid rated CDRs for VATable and nonVATable charging—as of course the customer needs to see it on the bills. However, I did not remember seeing the same flag on the prepaid CDRs. I went away, did my investigations, and found that the flag did not exist, and furthermore that in no downstream systems did this categorization take place. After this, the next steps were about proving and quantifying the issue.

I think this particular issue highlights an area of RA that is rarely, if ever, looked at. As an RA consultant, I spend the majority of my time looking at the processes and systems of both the network and IT functions. I tell them where they are going wrong and what they need to do, to avoid revenue leakage. Until this point in my career, I had never thought of looking at the department that most RA people and I sit in—finance.

Faulty Transmission

David Stuart

This question sees me returning to a standard RA issue.

At a fixed line operator, it was identified that approximately 50 percent of CDRs, from a single switch country, were not being received by the mediation systems collector. After a full investigation, the X25 transmission was found to be faulty, and hence only half of each CDR file was being transferred to the collector. The issue was identified without gaining access to the original CDR file from the switch or SS7 probe information. The issue had existed since the first day the switch had gone live, so call volumes were as expected.

What control would identify this issue?

* * *

This issue can be identified with one of the simplest RA controls to put in place. It was identified with a CDR sequence number checker. The majority of switches tag each CDR with a unique sequence number (a six- or eight-digit number). By simply comparing the difference between the first and last sequence numbers to the total CDRs within a file, along with the last sequence number in file A to the first sequence number in file B, you have a fool proof way of identifying whether you are receiving all data at the collector.

NOC versus RA

David Stuart

This question is not going to look at a revenue loss issue, but instead will focus on a common issue in an RA department.

I was working for a tier 1 fixed line operator and had just completed the development of an end-to-end CDR reconciliation. As with most of these types of solutions, the front end was also being used by the finance department to forecast monthly traffic volumes. An issue was raised by finance—they felt our count of calls coming off the switches were wrong, as figures from the network operations center (NOC) were much higher. The NOC had implemented an SS7 reporting tool, whereby they had probes attached to the entire transmission network that recorded every signal sent from each of the switches. These signals were aggregated into dummy CDRs from which they were able to identify the call volumes on any given day. Even when we removed the SS7 CDRs for which no switch CDR would have been created, the NOC still had figures 20 percent greater than those being seen by our solution. At this point we could have come to the conclusion that there must be a mass CDR suppression issue or CDR loss issue on the network; however, we did something else that proved our solution to be correct.

What would you do in this scenario?

* * *

To resolve this issue, we went for the most obvious resolution path and performed a CDR to SS7 reconciliation against a day's worth of traffic. After performing the reconciliation, we were still left with a large amount of SS7 records with billable duration that did not match to CDRs. We then started to analyze these SS7 records and the underlying customers, that is,

Had the customers produced any CDRs? — Yes
Were calls to the specific destinations producing CDRs? — Yes
Was it a specific duration issue? — No
Was there a unique flag on the SS7 record? — No
Did the SS7 records just originate from one switch? — No, but not the entire network

It was this last point that started to hone us in on the actual issue. All of the additional SS7 records came from a specific region of the country, which was served by eight switches. According to the company's engineers, there was nothing unique with this part of the network, so it was back to the analysis. At this point, we started to look at individual customer's calling profiles, and it was when we did this that we found the problem. We effectively had duplicate SS7 records, although they were not true duplicates in that the call times and durations were ever so slightly different. We had checked the data for duplicates right at the beginning of the analysis (as any good analyst will do); however, because the records were different, they were not flagged up to us.

Each pair of SS7 records was being recorded by different probes. The overall SS7 solution worked in such a way that only the originating probe generated the dummy SS7 CDR, so the fact that there were two records, both stating they were the originating probe, suggested something was wrong with the network, that is, double call routing. At this point, the issue was handed back to the engineers to work out what the cause was.

The network had originally operated an "A-Link" signaling network (signaling that goes down its own pipe via an SCP); however, the operator had moved to an "F-Link" signaling network (signaling is carried by the same pipe as the call) a few years prior to the investigation described above. The old A-Link signaling was supposed to have been fully decommissioned once the new F-Link was put in place; however, it transpired that in this region of the country, the decommissioning had somehow been missed. So what was happening was that for each call being made, the signaling was being sent across two pipes and hence two originating SS7 probe CDRs were being made—as there was a probe on every part of the signaling network. (I know—you thought they would have spotted the issue when installing the probes!!)

An SS7 to CDR reconciliation is one of the best pieces of analysis an RA department can perform—it allows us to verify the one piece of information we rely on for the majority of our other reconciliations (the switch CDR). Ultimately though, I hope the above issue has demonstrated that an SS7 record cannot be treated as definitive and does require the same levels of authentication as any other record produced by a network.

Fraudulent Prepaid Top-Ups

David Stuart

This question makes us return to the subject of fraud and its crossover into RA analysis.

While working for a mobile phone operator, it was identified that someone in the prepaid top-up card printing factory had been copying down the card's unique codes and topping up their own account. How would you identify this issue?

* * *

For this question, the root cause was one of many possibilities that were investigated. The actual issue was identified after validating data used for performing a prepaid balance audit.

A prepaid balance audit is a simple control whereby the following steps are performed:

Step 1: At a fixed point in time, record the total amount of credit held on your prepaid billing system.

Step 2: Record the total number of credits (top-ups) to the prepaid billing system for a given period (say 24 hours)—use an external source like the prepaid card database along with any other crediting mechanisms.

Step 3: Record the total amount of debits from prepaid accounts for the same period (as a negative number).

Step 4: At the end of the period, record the total credit held on the prepaid billing system.

Step 5: Perform the following reconciliation: 1 + 2 + 3 = 4.

Now, the above calculation did not identify the issue. However, by validating step 2 against cash received for the sales of top-up cards we could see that there was a general issue, where more credits were being applied to the prepaid billing system than those being sold.

Impossible Mission

David Stuart

For this question, I want to discuss one of the most annoying parts of working in RA: when data does not link. A typical example is a network to billing reconciliation of the ADSL product. Typically, a DSLAM will only contain information like DSLAM ID, port, slot, and maybe an IP address, whereas the billing system will generally have some sort of service identifier along with associated billing information. The data as such is unlinkable and therefore would appear nonreconcilable—what would you do in this scenario?

* * *

There are a few options available to us when data is not directly reconcilable. All of which hold benefits and pitfalls—so I will let you decide which is best.

1. Data Enrichment: In the majority of cases it will be possible to find a tertiary data set that can enrich your primary source to allow the reconciliation to take place. However, it is not just a simple case of updating one side of the reconciliation and proceeding as standard. Certain decisions need to be made:
   - Which data set do you enrich, that is, network, billing, or both?
   - How to treat records that cannot be enriched?
   - How to treat duplicate enrichments?
   - How to perform the reconciliation? It is technically now a three-way reconciliation, so should be dealt with accordingly.
2. High-Level Reconciliation: Even if data is not directly linkable at the individual record level, it is more than likely that it will be possible to reconcile at a grouped level, that is, total active network customers versus total billed

customers. However, if you perform reconciliation at this level there are certain risks/issues:

- – RA Equilibrium: You lose visibility of the individual issues and just see the net effect; that is, in its worse case you are overcharging 50% of customers and undercharging the other 50%,; however, due to a high-level reconciliation you see the net effect of perfect billing.
- – Revenue Loss/Gain: The high-level reconciliation shows a tangible error and in this case you cannot fix anything as you do not have detailed results, so you end up having to perform the detailed reconciliation anyway.

To avoid issues, the simple trick is to perform the reconciliation multiple times at different grouping levels; it will avoid the equilibrium issue and should enable you to isolate specific issues, thus avoiding an all-encompassing detailed reconciliation.

3. Data Source/Process Enhancement: Quite simply, get the original data sources enhanced by their business owners to include a common key. RA is not just about finding money; it is also about ensuring a good business process to reduce the risk levels against the company's revenues. To be clear, if there is no direct link between a customer's product and their bill, then there is an associated risk.

Summary

Ultimately the way I like to deal with these situations is by consolidating the three options above. First, I would perform the high-level reconciliation—this will generally find risks and issues that will allow you to justify the need for a business process enhancement by the data owners. To assist the data owners with the enhancement, I would then perform a one-off data enrichment exercise—identifying all records that can be easily updated and isolating the areas at risk. We eventually end up with a fully cleaned system that can be monitored with ease going forward.

No, No, NO!!

David Stuart

I am sure I am not the only person who has quoted the one-third rule as a "nice average" for at least the last 5 years, and I hope I am not the only person who knows why it cannot be used as a KPI—but why is it wrong to use 33 percent as a thumbnail rule or target for leakage recovery?

* * *

I think the easiest way to explain why a fixed target for revenue reclamation does not work is to first look at a couple of revenue loss scenarios per revenue stream.

One-Off System Faults

- A prepay billing system collapses for a period of time; during this period, all calls are free. Recovery is 0 percent.
- A postpay billing system collapses for a period of time; during this period, no calls are rated. The CDRs can be reprocessed and recovery is 100 percent.

Erroneous Reference Data

- A prepay billing system is rating calls at 1 pence instead of 10 pence. You cannot back bill. Recovery is 0 percent.
- A postpay rating engine is rating calls at 1 pence instead of 10 pence, depending on local regulations, duration of issue, company politics, etc. Recovery is between 0 and 100 percent.

From the examples above, it is clearly seen that there is a drastic difference between recovery in prepaid revenue loss and that of postpaid revenue loss. This is mainly due to the fact that prepaid recovery is rare (although not impossible); therefore, when defining recovery KPIs the first thing you need to do is separate your revenue streams.

What Is Recovery?

This seems like an easy/obvious answer; however, when you bring in the subject of opportunity loss, is it so clear? Well, can you recover an opportunity loss? Some people will say that because there is no billing error with an opportunity loss, just a service prevention, there is no revenue to recover. However, others will say that by enabling the service to be used, the revenues that are achieved going forward are the recovery. The debate then grows as follows: if forward billing is classified as recovery, do we therefore have to track all customer accounts forever, and if so how can we have KPIs? This debate can escalate exponentially and until there is a standard definition, any work on KPIs and other standard measures is pointless.

I personally believe that to make life simple, recovery can only relate to revenues already lost (the stuff in the past); for resolution of opportunity losses and the ongoing revenues from fixing a recovery, I would term these as benefits. (Please note that I am really watering things down here; there are many more benefits that need to be defined like opex and capex savings resulting from any resolution.)

Does Maturity Impact Recovery?

Of course it does! I am not going to write too much here, but quite simply, the level of maturity of a function dictates the types of revenue loss that is identified and, as we have already seen, "type" dictates recovery.

Summing Up

If we define recoverable leakages as being the moneys that should have been billed, but for some reason have not, then what we are looking at is something we really have no control over. Prepaid revenues generally cannot be recovered and postpaid revenues are so wrapped up in regulations and customer choice that again we have little or no control over reclaiming this money. What we do have control over are the benefits; every attempted recovery should have an associated benefit no matter what the type of loss.

I am sure that the headline of "Operators Recover Only a Third of Identified Revenue Leakages" is statistically about right if you exclude all of the unrecoverable loss types stated above. Ultimately though, in the case of the majority of African operators (99 percent prepaid), what you are talking about is less than 1 percent (financially) of the issues they will identify; therefore, the ratio is pretty much pointless.

If we want to look at RA as a whole, then a better study would be to look at all of the associated benefits of each revenue leakage prevented/recovered.

Low-Hanging Fruit

David Stuart

This question is more of a general discussion topic.

As a consultant I am often asked to find a quick win, either to demonstrate a team's capability or to get executive buy-in. To do this, I have a number of tricks/tactics that can often flush out a small amount of revenue loss.

How would you deal with this type of request?

* * *

There are a number of ways by which you can find quick RA wins within a communication provider; the following are just a few tips for picking the low hanging fruit:

1. Existing Reports: Most departments produce their own reports detailing customer numbers, system throughputs, revenues, and so on. Take these reports and compare the numbers, that is, NOC customer numbers versus finance customer numbers. Quite often you will find large differences that indicate that some sort of problem exists.

2. Interviews: Try and identify the key personnel who have manual processes that are revenue critical, that is, someone who manually inputs service orders. Interview the subject matter expert and you may find serious control weaknesses that indicate a potential revenue loss.
3. The Internet: This is my favorite and the easiest way to find revenue loss (fraud). Go online and search the Internet for free calls with your operator. You will find a number of chat sites where people swap tricks for getting free calls. Test out what they say as you may get lucky.
4. Customer Services: When customers get overcharged they will complain. These complaints will be logged in a CRM system and hopefully categorized. Try and identify the most common complaint. Obviously, what you will have is a revenue gain issue, but if the issue is the result of a system failure/problem, you will often find that the inverse issue exists whereby customers are being undercharged.

The above are just a few tricks that can be used, but the basic theme is simple; when faced with short timescales, do not try and source your own data. Just use what is already available.

Bibliography

Adams, D.N. (1999) *So Long, and Thanks for All the Fish*, Ballantine Books, New York.

Adams, E.K. and Willetts, K.J. (1996) *The Lean Communications Provider: Surviving the Shakeout through Service Management Excellence*, McGraw-Hill, New York.

Brown, M.G. (2004) *Get It, Set It, Move It, Prove It: 60 Ways To Get Real Results In Your Organization*, Productivity Press, New York.

Cartesian (2010) *Revenue Assurance Jargon Buster*, http://www.cartesian.com/resources/revenue-assurance-jargon-buster/ (accessed June 28, 2010).

Committee on Sponsoring Organizations of the Treadway Commission (COSO). (1992) *Internal Control—Integrated Framework*, American Institute of Certified Public Accountants, New York.

Dehiri, A. (2007) *Revenue Assurance: Identifying Leakage & Implementing Solutions*, Juniper Research, Basingstoke, Hampshire.

Dittberner (2006) *Telecom Risk Management: The Market for Revenue Assurance, Fraud, Credit, and Cost Management Solutions*, Dittberner Associates, Bethesda, MD.

Dunham, K.S. (ed.) (2008) *Revenue Assurance Overview (TR131)*, version 2.1, TM Forum, Morristown, NJ.

Forester, T. and Morrison, P. (1994) *Computer Ethics*, Second Edition, The Massachusetts Institute of Technology Press, Cambridge, MA.

Haig, M. (2003) *Brand Failures: The Truth about the 100 Biggest Branding Mistakes of All Time*, Kogan Page, London.

Kuhn, T.S. (1962) *The Structure of Scientific Revolutions*, University of Chicago Press, Chicago.

Lutz, W.D. (1989) *Life Under the Chief Doublespeak Officer*, http://dt.org/html/Doublespeak.html (accessed June 23, 2010).

Lutz, W.D. (1990) *Double-Speak: From Revenue Enhancement to Terminal Living – How Government, Business, Advertisers and Others Use Language to Deceive You*, HarperCollins, New York.

Magnus, P.D. (2005) *forall x: An Introduction to Formal Logic*, http://www.fecundity.com/logic (accessed June 23, 2010).

Mattison, R. (2005) *The Telco Revenue Assurance Handbook*, Lulu.com, Raleigh, NC.

Mattison, R. (2009) *The Revenue Assurance Standards*, Lulu.com, Raleigh, NC.

Priezkalns, E.R.A. (ed.) (2009) *Revenue Assurance Guidebook (GB941)*, version 2.2, TM Forum, Morristown, NJ.

Sevcik, P. (2010) *Comcast Usage Meter Accuracy*, NetForecast, Boston, MA.

Theil, S. (2008) "The Campus of the Future," *Newsweek*, August 9, 2008.

Wittgenstein, L. (1983) *Remarks on the Foundations of Mathematics*, The Massachusetts Institute of Technology Press, Cambridge, MA.

Womack, J.P. and Jones, D.T. (2003) *Lean Thinking: Banish Waste and Create Wealth in Your Corporation*, Second Edition, Free Press, New York.

Yelland, M. and Sherick, D. (2009) *Revenue Assurance for Service Providers*, Authors OnLine, Gamlingay, Bedfordshire.

Index